Other Related Titles

ISBN	AUTHOR	TITLE
0-07-911857-7	Ananthaswamy	*Data Communications Using Object-Oriented Design and C++*
0-07-005119-4	Baker	*Network Security: How to Plan For It and Achieve It*
0-07-005089-9	Baker	*Networking the Enterprise: How to Build Client/Server Systems That Work*
0-07-004194-6	Bates	*Disaster Recovery for LANS: A Planning and Action Guide*
0-07-004674-3	Bates	*Wireless Networked Communications: Concepts, Technology, and Implementation*
0-07-005076-7	Berson	*Client/Server Architecture*
0-07-005203-4	Berson/Anderson	*Sybase and Client/Server Computing*
0-07-005570-X	Black	*Network Management Standards: SNMP, CMOT, and OSI,* 2/e
0-07-005560-2	Black	*TCP/IP and Related Protocols,* 2/e
0-07-015839-8	Davis/McGuffin	*Wireless Local Area Networks: Technology, Issues, and Strategies*
0-07-020359-8	Feit	*SNMP: A Guide to Network Management*
0-07-020346-6	Feit	*TCP/IP: Architecture, Protocols, and Implementation*
0-07-021625-8	Fortier	*Handbook of LAN Technology,* 2/e
0-07-024842-7	Grinberg	*Computer/Telecom Integration: The SCAI Solution*
0-07-033727-6	Kapoor	*SNA: Architecture, Protocols, and Implementation*
0-07-035968-7	Kumar	*Broadband Communications: A Professional's Guide to ATM, Frame Relay, SMDS, SONET, and B-ISDN*
0-07-060362-6	McDysan/Spohn	*ATM: Theory and Applications*
0-07-042586-8	Minoli	*1st, 2nd, & Next Generation LANs*
0-07-042588-4	Minoli	*Imaging in Corporate Environments*
0-07-042591-4	Minoli/Vitella	*ATM & Cell Relay Service for Corporate Environments*
0-07-046461-8	Naugle	*Network Protocol Handbook*
0-07-046322-0	Nemzow	*FDDI Networking: Planning, Installation, and Management*
0-07-046321-2	Nemzow	*The Token-Ring Management Guide*
0-07-049663-3	Peterson	*TCP/IP Networking: A Guide to the IBM Environment*
0-07-051143-8	Ranade/Sackett	*Advanced SNA Networking: A Professional's Guide to VTAM/NCP*
0-07-054418-2	Sackett	*IBM's Token-Ring Networking Handbook*
0-07-057442-1	Simonds	*McGraw-Hill LAN Communications Handbook*
0-07-060360-X	Spohn	*Data Network Design*
0-07-067375-6	Vaughn	*Client/Server System Design and Implementation*

Network Security

Data and Voice Communications

Fred Simonds
Data Dynamics

McGraw-Hill
New York San Francisco Washington, D.C. Auckland Bogotá
Caracas Lisbon London Madrid Mexico City Milan
Montreal New Delhi San Juan Singapore
Sydney Tokyo Toronto

Library of Congress Cataloging-in-Publication Data

Simonds, Fred.
Network security :/ data and voice communications / Fred Simonds.
p. cm. — (McGraw-Hill series on computer communications)
Includes index.
ISBN 0-07-057639-4 (hc). — ISBN 0-07-057634-3 (pc)
1. Computer security. 2. Computer networks—Security measures.
I. Title. II. Series.
QA76.9.A25.S35 1996
005.8—dc20 95-30723
CIP

McGraw-Hill

A Division of The ***McGraw-Hill*** *Companies*

1 2 3 4 5 6 7 8 9 0 DOC/DOC 9 0 0 9 8 7 6 5 (HC)
1 2 3 4 5 6 7 8 9 0 DOC/DOC 9 0 0 9 8 7 6 5 (PBK)

ISBN 0-07-057639-4(HC)
ISBN 0-07-057634-3 (PBK)

The sponsoring editor for this book was Jerry Papke, the editing supervisor was David E. Fogarty, and the production supervisor was Donald Schmidt. It was set in Century Schoolbook by Dina E. John of McGraw-Hill's Professional Book Group composition unit.

Printed and bound by R. R. Donnelley & Sons Company.

This book is printed on acid-free paper.

For my wife, Laura,
and my two daughters,
Jennifer and Rachel

Contents

Preface

Today, networking is getting a lot of attention. There is talk of the information superhighway and cyberspace. The Internet is on everyone's lips. One is out of step with the times if they lack an e-mail address. Where did all this interest come from?

LANs surely stimulated networking. Fiber optics lowered the cost and improved the quality of networking forever. Hardware costs are in the cellar, making complex technologies economic.

But these overlie the real reason for networking's popularity. The bottom line is that people want the quick response and service that networking offers. Networking and service to customers go hand in hand. We expect ATM machines to be always on-line and our credit cards to be validated in a moment or two. As consumers, we expect instant service. So nothing less than an on-line network will suffice. Quick response and adaptability have become business bywords and a staple of commerce in the 1990s.

Networking goes right to the consumer in the shape of volksnetworks such as America Online, Prodigy, CompuServe, and the Internet. These networks bring e-mail to the masses and information to our doorsteps. Our service expectations build ever higher.

All these fine things have done something else too: They have exposed our activities to those who would cause us mischief at the least and real harm, both monetary and bodily, at the worst. Security has taken on new urgency and is a gating item for new services such as doing business on the Internet. Traditional networks such as voice and cellular networks are being retrofitted to add security in the face of serious losses.

Security: Solving the People Problem

This is a technical book, but it remembers that security is always a *people* problem. We simply use technical measures to protect ourselves from sociopaths who want to get for nothing what we have worked to create. The form of the problem may be new, but the cause lies deep inside us and is as old as mankind. It is of paramount importance that amid all the technical minutiae that we retain the essential perspective that machines do not steal or betray, *people* do.

Another side of the people problem is that honest users are not sufficiently security-conscious. A continuing campaign is necessary to keep security in the forefront of every user's mind. Management sets the tone by their example and by the way they support the security effort with their time and their organization's money.

The Need for a Global Policy

Two themes recur in this book again and again. The most essential is to have a security *policy*. Until and unless you have a realistic policy in place, it is impossible to build workable security protection. Distilling a practical policy, one that balances access, protection, and cost, is an iterative effort, and one that should involve all levels of your organization. This book gives you specific guidance on how to build an effective policy.

The second theme is that the policy and its subsequent implementation must be *global*. By that I mean consistent, and of even strength throughout the organization. Global consistency denies attackers the single weak spot they need. Serious attackers are patient, persistent, and intelligent. They have a lot to gain, little to lose, and time is on their side. Some of them enjoy six-figure incomes. This book is intended to help keep them off your payroll.

Constant vigilance is the price we pay for security. Unfortunately, being constantly vigilant is not something human beings do well, so we enlist our machines to help us. This book shows you ways to do that. In the words of one of my mentors, "Your security is only as good as the eyes on your network." It took me a while to learn that he meant electronic eyes, eyes that never sleep but do blink occasionally.

Internet References

This book contains many references to Internet information resources, both in the text and at the ends of the chapters. It was impossible to include or even distill all the information available into a single book. Better to state the principles here and allow you to seek out detailed information to the level you need as you need it.

Internet references help to keep this book current. Threats constantly evolve. New threats will be discovered, analyzed and, countered through dynamic forums such as alt.security and comp.security.misc. We want you to know where to get the latest information, how to keep an eye on threats that face us today and be forewarned about those that will raise their ugly heads tomorrow. Even better, we urge you to become a contributor to these forums and help us as we try to help you.

For Whom Is This Book Written?

This book is intended for the practicing security professional or for the individual who is tasked with implementing and/or maintaining network security. Unlike some other works, this book is distinctly practical and nontheoretical. My experience in teaching hundreds of seminars is that most people prefer to know the strengths, weaknesses, future outlook, and uses of a technology rather than become immersed in a lot of technical detail. Information that helps make good decisions is a wise use of time; all else is gilding the lily. The former approach is taken here. The security principles put forward in this book are broadly applicable, and work just as well for those tasked with nonnetwork security responsibilities.

Organization at a Glance

The book is organized like this:

Security threats. Chapter 1 profiles the categories of attackers and what motivates them. The objective is to look at our network as they do, and in so doing detect weak spots and plug them. Chapter 2 discusses computer viruses in both a networked and nonnetwork setting. It discusses viral forms, signs of infection, and how to prevent an infection or cure one if your system becomes infected.

Security tools. In Chapter 3 we itemize and compare the security standards that permeate the industry with a view toward their security, expected longevity, and utility. Authentication and encryption are closely related and indeed overlap in some areas. The similarities and differences are compared and contrasted in Chapters 4 and 5. Controlling access is a key part of security administration. Access control principles and applications are discussed in Chapter 6.

System security. These chapters examine security in the context of specific systems. LAN security is detailed in Chapter 7, while operating system and network operating system holes occupy Chapter 8. Leaks from interconnected LANs and WANs are analyzed and solutions suggested in Chapter 9.

Firewalls. A full discussion of firewalls, security problems, and solutions in TCP/IP systems and firewall configurations to meet a variety of requirements are discussed in Chapter 10. Security testing tools such as SATAN and Pingware are also analyzed. Many useful information resources are noted within the chapter and more are in a list at its end. The listings include where to find copies of SATAN and sources of free, yet effective, firewall software.

Voice systems. There is very little written about telephone fraud, yet the annual losses run into billions of dollars. This chapter points out forms of voice fraud, how to detect it using your own systems and those offered by the carriers, and how to shield your system from intruders.

Security development and management. How much security is enough? Chapter 12 presents guidelines to help you find the right balance of security. As mentioned above, Chapter 13 tells you how to create that all-important network security policy. So where to begin? "A journey of a thousand miles begins with a single step." In Chapter 14, we help you take important steps toward securing your network. It will surely be an interesting passage.

FRED SIMONDS
fredsi@ibm.net

Acknowledgments

While the author's name may go on the cover, there are a lot of people who help him or her get there. I cannot possibly know all the people who worked behind the scenes to make this book a reality. I must thank you all anonymously.

Jay Ranade, Series Advisor for the Series on Computer Communications, examined my initial proposal and presented it to McGraw-Hill. A classic overachiever, I use his experience with over 125 titles in all as a reality check. Jerry Papke, the sponsoring editor for this work, kept tabs on its development and made a number of useful suggestions that I took to heart and incorporated into the book. The editing manager is the conductor of this orchestra of people, including me. David Fogarty of McGraw-Hill kept everyone in tune so that we achieved some kind of harmony.

My friend and well-known colleague Bud Bates reviewed the work technically. Taking the reader's perspective, he carefully reviewed the manuscript. He suggested a number of changes and enhancements, all of which I incorporated into the book. Proactive as ever, he sent me some unique, hard-to-find material for the chapter on voice systems that strengthened it considerably. However, any errors are my responsibility and your comments are welcomed via e-mail to fredsi@ibm.net.

I would not be doing this book with McGraw-Hill were it not for Dr. Sidnie Feit's helping hand with my first volume, *The McGraw-Hill LAN Communications Handbook*. Though she was not involved in this new project, I have applied and tried to refine the many lessons I learned from her.

My family has endured my many hours at the computer and countless away-from-home seminars. Their patience and encouragement is something I acknowledge every day.

FRED SIMONDS

Chapter

1

Where Do Security Threats Come From?

Objectives

At the end of this chapter you will be able to:

- List intentional and unintentional threats to information
- Catalog sources of unintentional leaks and how to correct them
- Recognize internal and external threats to your organization
- Itemize internal threat sources and how to neutralize them
- Categorize the levels at which "hacking" takes place
- Describe the hacker profile
- List the basic steps required to make it hard for hackers to invade your system
- Discuss commercial espionage, and its possible effect on your organization
- Describe the scope, capabilities, and effect of government-sponsored espionage
- Explain how encryption key escrow systems work, and where they are used
- Discuss transborder encryption issues
- Recognize danger signs an employee sends off and organizational red flags
- See how vendors can be a security risk, and how to prevent it

- Evaluate the extent of the virus threat as a prologue to Chapter 2

Prologue

You work for an organization. You produce goods and/or services. They have value in the marketplace, or you would not be doing it. Your organization took the risks, did the research, and made the investments. You are entitled to the fruits of your labor.

Unfortunately, there is always a human element eager to take advantage of the work done by you and others, yet without the customary quid pro quo: payment for services rendered or for product provided.

Direct theft of products or services is a simple motivation. More complex is the theft of information for a variety of nefarious purposes: to improve a competitor's position; to further a political aim; to deny you timely access to correct information that you need to survive; to give the perpetrator an ego high.

Security is entirely a people issue. It is *people* who cause security problems, not machines or systems. Fortunately, our high-technology tools can also serve to thwart people who would do us harm, just the way we lock our doors against intruders. That security is a people problem must be borne in mind as an underlying constant theme in any discussion of security.

Even well-intentioned people are a weak point. Well-meaning people write down passwords or let others use theirs. While we try to use "social engineering" internally to modify behavior so that people will keep security in mind, outsiders try to use the same principle of social engineering to talk users out of passwords. In a sense, they are taking advantage of our incomplete or ineffective social engineering, trying to replace it with their own.

Now we are mentally prepared to begin to think like our adversaries. If we can put ourselves in their place, then we can do some role-playing and begin to view our security precautions not just from the inside, but from the outside. Such a dual perspective is essential to effective security management.

What Can People Do to Our Information?

- They can *destroy:*

  Information

  Executable programs

  Operating systems

Other computing or system resources

Destruction includes both logical damage (to the system involved) and/or physical damage to the equipment itself. More subtly and short of destruction, tamperers can alter any of the above, presenting the appearance of legitimacy though the reality is false.

- They can *steal:*

  Information

  Service

  Hardware

  Software

Theft includes simply taking information not belonging to them, whether acted on or not. It includes the unauthorized use of system resources and is exemplified by the theft of telephone service. Theft plainly includes removal (such as deletion of information) as well as physical transport off the premises.

- They can *disclose:*

  Information to which they have no right or need

  Information to unauthorized personnel for personal gain or for other motivations

  How to let others use resources not intended for outside use

- They can cause *service disruptions or interruptions:*

  By causing physical or logical damage to the system

  By improperly denying access to legitimate users

It is important to note that motivation is not always relevant. Any of the above losses can be caused accidentally. Furthermore, the above can be caused as acts of omission (such as failing to do a backup) or commission (such as deliberately causing a malfunction).

Classifying Threat Sources

A threat is first classified as originating from outside or inside the organization. From there, we proceed to more specific subcategories:

External threats include:

- Hackers
- Commercial espionage
- Government-sanctioned espionage
- Vendors
- Former employees

Internal threats include:

- Disgruntled employees
- Unintentional losses or security breaches
- Hackers

The bulk of the case evidence shows that it is the *internal* and not the *external* threat that must concern us the most. Kenneth Weiss, the founder and chairman of Security Dynamics, says that 78 percent of computer abuse originates with a company's own employees. Even so, instances of external interference are growing.

Hackers

Originally, the term meant someone who made furniture with an axe. Today, the term *hacker* is used one way by the literati of the computer community and in another less flattering way by the public at large. In the computing community, a hacker is a talented programmer whose own skill compensates for a lack of software development tools. These skills are usually acquired the hard way, through trial-and-error experience over time.

This is not the definition attributed to hackers by the general public. The common definition of a hacker is one who gains access to a system where he or she does not belong.[1] Within the computing community, malicious hackers are distinguished from the benign species by referring to the former as *password hackers, network hackers,* or simply *crackers*. People who break copy protection software are also called crackers.

For nonmalicious hackers, the motivation is not material gain: it is the challenge, specifically to stretch their talents or to find ways to do things thought to be impossible; to stretch the limits of the technology using their ingenuity and creativity. This does not imply even in the smallest way that they are doing anything illegal. Hacking in this sense gives these people an outlet for their tremendous intellectual energy. Taken as a whole, the notion that hackers are all malicious is far more wrong than right. In fact, to be called a hacker on the Internet's USENET is a compliment. Calling someone a cracker is to invite trouble. Mixing the terms unknowingly is to invite bigger trouble still.

The vindications that "good" hackers apply to people crawling around in systems where they have no right to be boil down to self-serving rationalizations. Many hackers argue that information (except that which is personal and confidential) should be freely accessible. The idea that information cannot be owned and should be shared across all society is a tenet that first emerged in the Enlightenment and has become a long-standing principle of Western thought. By this argument, the establishment of artificial security restrictions holds back the advancement of technology and society at large. Society's evolution into the Information Age, in which information itself is a source of wealth, implies that information, like wealth, can be treated as ownable property.[2]

Indeed, vendors who have insisted on taking such a proprietary approach toward their technology have generally fared worse over time than those who have shared their technology with the public. This has proved true with IBM personal computers versus Apple, the generic Transmission Control Protocol/Internet Protocol (TCP/IP) versus Systems Network Architecture (SNA) and DECnet architectures, and Novell NetWare versus Banyan VINES, to name just a few examples.

One hacker argued that if system managers were not so sloppy, then hackers would not be able to gain entry in the first place. He estimated his success rate at 10 to 15 percent with a self-imposed limit of one hour per target system. This rationalization is tantamount to saying that if you leave your gold watch on a picnic table in the park, then I have the right to take it. While you as the owner have a duty to protect your property, the fact that you do not fulfill your duty does not give me the right to take your property.

Some hackers wish that system managers would stop seeing them as ever-malicious and instead use hackers' talents to help secure the systems they enter. The idea of using hackers as part of "tiger teams" has merit. This has been done in a number of cases, with positive results. In an effort to find a quid pro quo, hackers have suggested trading access to these generally sophisticated systems in exchange for their help in nailing down security. This idea has been received coolly by the security community; it seems a little like asking the fox to guard the chickens. Some hackers would say that this is precisely the problem: they are perceived as inherently evil. It is clear that some hackers are morally affronted that the fine systems they invade are sometimes treated with sloppiness, disrespect, and even incompetence by their owners or administrators.

Another creative idea is for hackers to attempt to gain entry to a system and then leave their "calling card" somewhere in its innards to announce their presence and inform the system manager how they got in. Unfortunately, this idea is hopelessly naive in a corporate environment where appearances mean more than reality. Applied to this instance, such a scenario would soon have the system manager looking for work.

Part of the hackers' image problem is the overwhelmingly negative hype pumped out by a sometimes unwitting, frequently uninformed, and often self-serving press. Heavy-handed, knee-jerk government policy and enforcement, based on the classic idea of punishment to fit the crime, is regarded by some security professionals and criminologists such as Jim Thomas and Gordon Meyer as a waste of precious law enforcement resources. Again, many law enforcement people hold the simplistic view that hacking is malicious by definition.

Even if it is granted that invasive hacking is not inherently damaging, the owner of a machine has the absolute right to determine who will use it, right down to the last machine cycle. Furthermore, it is not the purpose of a machine's owner, at least not in the business community, to advance technology for its own sake. Such an owner is not interested in philosophical dialogue about knowledge dissemination in the Information Age; they are interested in advancing their organization's mission, often the acquisition of monetary wealth. Hackers do not have the right to utilize other people's machines to achieve social objectives they deem desirable without the knowledge, much less the permission, of the machine's owner.

The argument that a nonmalicious hacker does no damage and so legitimizes entry without permission is specious for the same reasons. The very act of intrusion is disruptive, even if undiscovered. If discovered, backups are often invoked and other actions taken because the system manager has no way of knowing what, if anything, was taken, copied, damaged, or planted. The simple fact that most hackers are not malicious is totally insufficient to allay the concerns of an increasingly paranoid public. Not knowing what the hacker did only fans those mistrustful flames. In short, there is no way to know or anticipate the point at which innocuous hackers become crackers, if ever, and no one to regulate their activities except themselves.

As Fig. 1.1 shows, cracking is a progressive process that often starts innocently and proceeds in phases to breaches of greater and greater seriousness. Sometimes, usually in the beginning, hackers experiment, test the waters, and hone their programming skills. Once they gain entry into a system where they do not belong, these people become *crackers:* they crack a system's security. The FBI calls this first phase the *intrusion phase.* Many crackers start by breaking into a victim's telephone system. Once in, they can go anywhere and may be untraceable.

Crackers can have a legitimate purpose. For example, a person hired to test security measures is a lawful cracker. "Tiger teams," originally conceived to test nuclear facility security, are an accepted security technique.

Some crackers progress to the second phase: they leave a whim-

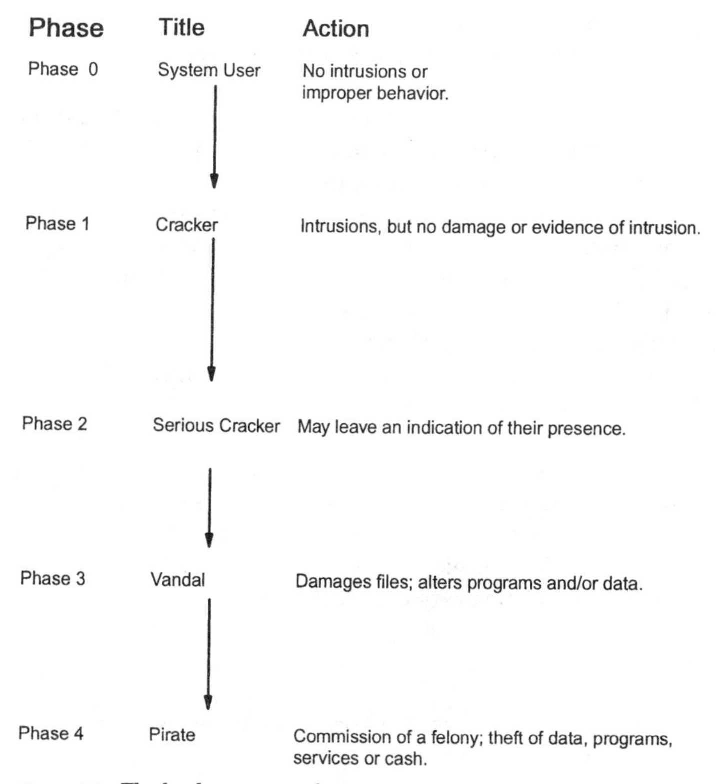

Figure 1.1 The hacker progression.

sical message, or a nondamaging virus, just to let people know they were there. This small but portentous step illustrates how insidious the progression is. A series of small steps leads the casual experimenter toward greater and greater mischief.

The third phase turns the cracker into a *vandal:* vandals damage files, alter programs, and the like. At this point the organization begins to suffer some kind of loss.

In the fourth and final phase, the now-established vandal commits outright felony. This often involves theft of data, programs, cash, or other items of real value. The lure of unlimited credit potential in a credit card system, for instance, may prove too enticing to resist. Upon entering the fourth phase, the vandal has become a *pirate*.

The phase-by-phase progression has all the earmarks of an addiction. That is what it is—a *behavioral* addiction. Like many addictions, the recidivism rate is very high.

Some hackers skip right to the final phase and simply become criminals, out to steal. Others are self-exiled social outcasts. But many are young people who are testing themselves in a primordial soup of yet-unformed morals. Like many youths, most fail to understand the consequences of actions they think of as pranks.

In the words of Special Agent Jack Lewis of the U.S. Secret Service: "It used to be that people would consider [hacking] an intellectual prank. But now, we're seeing more and more people do it for baser motives. We've seen a tremendous increase in computer crime."

One major reason for such attacks is that few hackers get caught. A California bank thief—never captured—got away with $25 million. Many others are never apprehended either. Understandably, embarrassed victims do not care to publicize their losses. As a result, annual loss figures vary widely from $500 million to billions. No one really knows.

Many of these attacks are against long-haul carriers such as AT&T, MCI, and Sprint. Attacks on corporate private branch exchanges are also growing. In effect, successful attacks against carriers and corporations finance hackers' activities. Using stolen telephone service, hackers seek entrance into other systems, compounding their crime at corporate expense.

For other kinds of hackers, corporate espionage in return for money is their objective. There are plenty of examples:

- Companies are approached with offers to steal competitive information for money.

- A hacker in jail is approached by other white-collar criminals to use his skills once free.
- A group from former West Germany penetrated computers attached to the Internet. Their objective was money and drugs. Their client was the former East German secret police.
- Allegations that hackers based in Detroit are being used by industrial spies instigated a federal investigation. A phantom group of French hacker mercenaries known as QSD keeps the authorities guessing.
- Hacker Leonard Rose pleaded guilty in March, 1992 to wire-fraud charges. According to prosecutors, he had disseminated a proprietary Unix code taken from a former AT&T employee.

For these people, no innocent claim of challenge or intellectual curiosity can possibly pertain. They are thieves, pure and simple.

A hacker profile?

The usual stereotype of a hacker is a moneyed, antisocial, loner, computer nerd with low self-esteem and a high IQ. Unfortunately, the reality is nowhere close. While there is no narrow profile into which a hacker neatly fits, there are some broad descriptors:

- A hacker resents authority. Hacking has appeal because it is forbidden. Even better, there is often no single or identifiable victim. Rationalizations come easily.
- The hacker seeks challenges of ever-increasing difficulty.
- Often personally ineffectual, hacking offers a way to obtain power and control.
- The hacker is the kind of person who seeks to avoid taking responsibility for his or her errors.
- As a corollary to the above, hackers dislike taking risk.
- A hacker has indistinct, unformed, immature, and therefore fuzzy definitions of right and wrong.
- For example, some hackers believe they have a "right" to visit other people's systems.
- Hackers have long memories. Emotionally immature, they may harbor a grudge long after an incident has been forgotten by others.

- Some hackers live in a fantasy world that bears little relation to reality.

The hacker ethos

You have surely read stories about hackers who became security consultants after having been "reformed" in some way. This seemingly positive practice sends a very clear message to hackers: even if you do evil, you can change hats and become a good person. After the penalty comes the reward. This is not the desired disincentive to would-be hackers!

Young hackers get what they crave: recognition, notoriety, and status. Stories are written about them in the newspaper. All the publicity feeds the myth that these people are modern-day Robin Hoods with PCs.

In truth, many of today's computer professionals had some experience with hacking early in their careers, including the founders of Apple Computer who made infamous "blue boxes" that generated tones to disable telephone company billing equipment.

Anyone with a known background in hacking should not be hired, either as consultant or employee. To do otherwise is to perpetuate the notion that what hackers do or did is somehow noble or socially acceptable. It is neither.

In a positive vein, some employers make employees sign an agreement indicating what areas of the computer system(s) they may access and those that are off limits. However minimal, the employer gains some protection as a deterrent and perhaps as evidence of a contract that was breached should the matter wind up in court.

Thwarting a hacker attack

An ounce of prevention is worth a pound of cure:

- *Warn.* System users, including inbound callers using modems, should be warned that the system is private and for organizational use only. This is often done as a login screen banner.
- *Check passwords.* Review passwords for obvious or weak ones. Some systems will not accept a password that is too short (e.g., less than five characters) or weak (e.g., repetitive such as 111111).
- *Promote.* Post news stories about hackers and security breaches. This sends a twofold message: it keeps security up front in the user's mind and reinforces the fact that it can happen.

- *Test.* Test published and known breaches in your own system. Do you have procedures in place that will prevent them?
- *Monitor.* Track your long-distance telephone traffic. A sudden increase may mean a hacker is lurking in the system.
- *Keep your eyes—and mind—open.* It is tempting not to look for fear of what unpleasantries one might find.

Figure 1.2 illustrates the security layers typical to an organization.

Commercial Espionage

It almost sounds too James Bond–like to be so, but illegal commercial espionage has been around a long time, especially in bitterly competitive industries such as the automobile industry. Today, *every* business is equally competitive.

Clearly, much of this espionage comes from that competitive pressure. One of the author's clients was involved in oil exploration. Seismic data from test platforms at sea were radioed back to the company's research laboratory. The information was used in

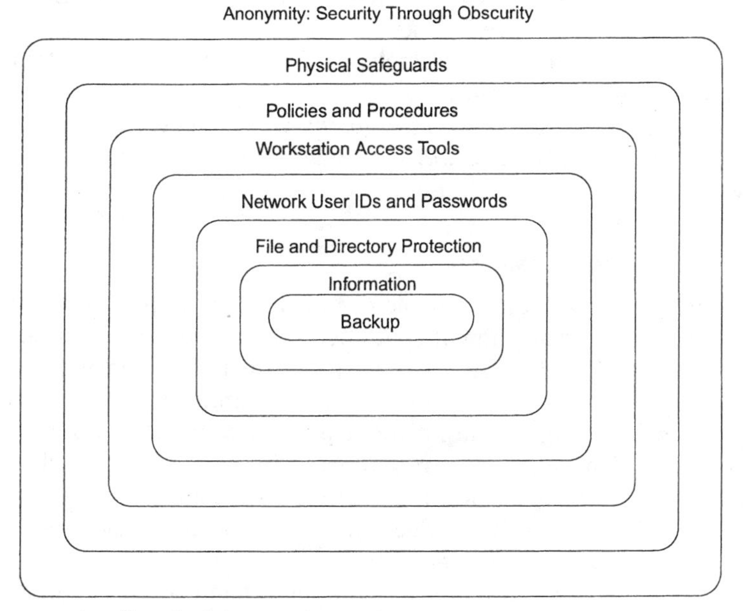

Figure 1.2 Security by successive attrition.

making decisions about whether to bid for undersea oil leases, and if so, how much to offer. For a time, the procedure worked well. But then the client's company noticed that a competitor was bidding for much the same leases and slightly higher. The competitor began to win a lot.

The client installed encryption equipment to scramble data going to and coming from the test platform. Eventually the competitor's winning pattern disappeared. No one will ever know whether the encryption equipment really made the difference, or if the competitor was really listening in. But if you believe in cause and effect, then there was commercial espionage in progress.

Using information so monitored is illegal, because under federal law, radio signals received by those other than the intended recipient may not be used by them. Such transmissions are considered confidential. In practice, the law is broken every time a cellular phone conversation is taped and every time a police radio scanner is overheard, sending a reporter to the scene of an accident or crime. Obviously, the law is rarely enforced and in some sense is unenforceable.

There are many other cases of companies spying on other companies, using paid professionals. In one such case, the phone lines of the president and CFO of a company were tapped—in their own phone closet. Data lines, always so neatly marked by the telephone company, were just as neatly tapped.

In a highly publicized case, Hitachi paid corporate spies to steal IBM secrets. They were caught.

Commercial spies use *passive monitoring* with sensitive receivers in a van on the street to pick up electromagnetic signals from video displays. Someone's paying them!

Commercial intelligence gathering

Commercial gathering and analysis of information is not illegal in itself, but it is easy to see how it could become so. *Competitive intelligence* consists of gathering all the legally available information about an organization. Piecing it all together, analysts review the result, perhaps giving a competitor a picture of what's going on inside the organization. This is no different in principle from what the CIA does—analysis—when it collects information about a foreign activity. As long as the collection process is legal, the use of competitive intelligence is perfectly legal.

It is not hard to see how an unscrupulous business intelligence gatherer could ally him- or herself with a hacker to develop an information pipeline. The gatherer could insulate him- or herself from the hacker by using a go-between or by pretending to be unaware of the illegal means by which the information was obtained.

Intelligence gathering need not be passive. Reports from field people, rumors, and trade paper reports all give a certain picture. In fact, a model built this way is generally far more accurate than the limited picture a tiny piece of stolen information can provide.

In sum, the legal activity called *commercial intelligence* (CI) gathering and analysis is separated from commercial espionage only because CI practitioners are supposedly self-bound by a self-regulating code of ethics. Clearly, this is not a particularly strong method of regulation. CI practitioners cross the line when hackers start feeding them information. It is hard to see how a CI practitioner could resist so tempting an apple.

Foreign Government Industrial Espionage

With the end of the Cold War, there are a lot of unemployed spies out there. How does a spy find a market for his or her unique skills when confronted with a new world order in which those skills are no longer in demand?

At the end of the Cold War the FBI established a New Threats Security List. The first item on their list is economic espionage. Former Soviets themselves are no longer perceived as the main or only threat: there are cases of Japanese, German, and French government-sponsored intelligence gathering on American companies through hacking.

For example, Nippon Telephone and Telegraph (NTT) cooperates with Japanese companies who want to enter American companies' databases in Japan. The Ministry for International Trade and Industry (MITI), uses NTT to decode faxes. That information is passed along to Japanese industry at cost.

This practice (called "fax-spionage") is ludicrously simple: fax formats are all standard. As a result, Hewlett-Packard and Boeing (among others) do not use facsimile to Japan. They use the mail or overnight express.

Former West Germany launched Project Rahab in 1987 to breach U.S. companies' databases around the world. Reportedly the German agents breached DuPont's security and valuable information was passed to German chemical companies. This was and is nothing less than *state-sponsored espionage by a friendly government.*

In France, Pierre Marion set up a section of French intelligence in 1981 to spy on American firms. These intelligence agents hacked

U.S. firms' databases. They were even foolish enough to brag about it, prompting an embarrassed apology from the French government.

You might think that encryption would solve the problem. However, some countries will not allow transborder encryption of information. The Swiss will not allow the American-originated Data Encryption Standard (DES) to be used. In some countries, such as Japan, Germany, France, and South Korea, a company must give copies of the encryption keys to the government. The government can—and does—tap in whenever it wishes.

The U.S. government is not very tough on industrial espionage. Generally, if the information is not a military secret, prosecution is impossible. Nor does the United States restrict the use of private keys or encryption of data going abroad.

In France, selling information of economic or technical value can lead to jail. There is some sentiment that the United States should make foreign companies post their keys with us, if we must post our keys with their governments. Our lopsided policy has placed us at a definite economic disadvantage, to say the least.

In the United States, it is no secret that the National Security Agency (NSA) monitors traffic in and out of the United States, with the cooperation of long-distance carriers such as AT&T. NSA has developed sophisticated algorithms to pick out information of interest and can break many encryption schemes. However, information so obtained is used by the government only and is not generally disseminated to private industry. Once again our policy is out of balance when compared to the way other countries manage such activities.

In the United States, the use of a key escrow system is now under active consideration. Designated *Capstone,* the user of an encryption system would deliver a copy of the encryption key to the federal government. Only under court order could the key be used to monitor electronic communication. We will discuss this in greater depth in Chapter 5.[3]

Internal Threats

To keep our perspective, we must recall that more threats come from inside than outside, primarily because employees already have access. The internal threat is all the more unfortunate because knowing this makes us wary even of people who really are on our side.

Clearly, such misgivings do not make for good employee relations. Employees may feel suspect from the day they start work. If they are required to sign an employee security agreement, their sense of paranoia heightens. A faint air of mutual suspicion is the result. In a time of employee empowerment, security concerns dampen enthusiasm and send an implicit and conflicting message: "We may say we empower you, but we really only trust you just so far."

Most often, the internal threat manifests itself in the form of a disgruntled employee seeking revenge. An ex-employee, especially if discharged for cause and seeking retribution, is the next most common motivation.

Since employees know their way around, they often have an idea of where the company is vulnerable. Furthermore, they can leave "logic bombs" and other software bugs that can take months or even years to go off. On the appointed day, for instance, a system's entire set of files is erased. The perpetrator has left no clue and is long gone.

Virtually all information systems (IS) people know by now that once an employee has given notice or been terminated, all system access should thereafter be denied. While there is no way to know what the employee did before resigning, there should be an audit trail to find out where they have been lately. This line of inquiry should be pursued if the employee is leaving under less than friendly circumstances.

Some departments become too aggressive about turning off an about-to-be ex-employee's access. Here, timing is important. If IS shuts off an employee's access to the computer or the telecommunications department turns off the employee's telephone or voice mail access, then the soon-to-be ex-employee discovers the hard (and improper) way that he or she is on the way out. Between that realization and the actual termination, such a person may be able to promulgate other mischief with no supervision or oversight.

The proper way is to turn these functions off during the termination meeting or immediately thereafter. If access is still needed, then it should be supervised by another employee.

Other internal problems

Some employees occupy uniquely powerful positions within the MIS management hierarchy. Specifically, these are local area network (LAN) administrators and database administrators.

LAN administrators manage one or more LANs, often including hundreds of users on several servers. They manage password assignments and files, backups, security, and application access. They are in a uniquely powerful position, with direct access to network server(s).

A malicious LAN administrator is in a position to do a lot of harm, ranging from erasing passwords to killing applications to damaging files beyond recovery. The best protection against a malicious administrator is to divide and conquer. Do not create a LAN czar. Rather, divide the responsibility among several administrators. Furthermore, make them work together so that collusion is less likely. Keep backups available so that if the worst happens, you are covered.

A few firms have gone so far as to bond their LAN and database administrators. But monetary reimbursement will not make up for the lost time and lost data that can result if an administrator gone bad is not promptly removed.

If anything, database administrators (or DBAs) are even more powerful. The DBA maintains the database so that it can serve all authorized users. These tasks include installing database software, supporting applications, handling documentation, and maintaining security. The risk would be reduced if these responsibilities were spread around. But too often they are not. When they are concentrated in one person, that person becomes a czar.

Czars can change the database for their own benefit. They can give themselves privileges that allow access to areas where DBAs do not belong. They can allow a dishonest co-conspirator to gain access. DBAs can falsify transactions (e.g., credit card authorizations) or modify them. They can even control the backup process so that the backup file will be corrupted or the backup information is changed before appearing on line.

The best (or worst) real-life example is that of the Equity Funding insurance scandal. The chief actuarial officer, who was in effect the DBA, created certain insurance policy records and hid others. Because the data processing responsibilities were not divided among several people, the officer was able to manipulate the records. (There were also clear indications that other managers colluded as well in the fraud.)

It is common to believe that sharing responsibility is a strong deterrent to collusion. While this is so, sharing of responsibility is by no means a magic bullet and is far less effective than common sense would indicate.

Employee and organization danger signals[4]

Suspect trouble when the employee:

- Plainly lives beyond his or her means

- Gambles excessively or has possible drug and/or alcohol problems
- Resents superiors and may show job frustration
- Has low self-esteem: excessively seeks power, status, and influence
- Has been a job-hopper
- Has a shaky financial and/or credit history
- Does not take vacations (so others cannot see what the employee is doing; sometimes there may be a loss of ill-gotten income)
- Is hard-pressed for productivity performance

At the same time, trouble may be triggered when the organization:

- Is in a downturn
- Keeps quiet about employee irregularities
- Inadequately screens employees in sensitive positions
- Does not compel executives to make financial disclosures
- Trusts key employees too much
- Has poor compensation practices
- Is always in a state of crisis
- Has inadequate internal audit, control, and accounting practices
- Has inadequate training programs

Unintentional leaks

Unintentional leaks fall into three areas:

- An "eyeball" leak: someone observes a printout or CRT screen of information they should not see. Eyeball leaks are also referred to as "shoulder surfing." Eyeball leaks are a special problem with fax machines.
- A person discovers an unsought way into a system.
- The focus of security on one area leaves another vulnerable.

Eyeball leaks. Eyeball leaks are easily handled, but call for constant vigilance. Some suggestions are as follows:

- Use screen blankers after a few seconds if the terminal is inactive. Alternatively, set the screen saver timeout to a short period.

For extended inactivity, automatically log the user out of applications (and a network if logged in).

- If using paper fax:

  Cover all faxes to be sent with a cover sheet.

  Cover all faxes received the same way.

  Do not permit messages on cover sheets.

  Delegate fax transmission and delivery to a secretary or clerk, and no one else. Have him or her log incoming and outgoing faxes.

  The best solution is to circumvent the whole problem by using a fax server to send and receive faxes. This can be a productivity boost, too, as some 70 percent of faxes exist in electronic form to begin with. It drains productivity to have to convert to paper first!

Falling through the trap door. In the course of working with a system for awhile, users may well find ways into an area where they do not belong. They did not go looking for the back door, but they found one anyway.

In the best case, they will bring it to the attention of the appropriate administrator. In the worst case they will exploit it and eventually may become hackers. Between these extremes, most observe it and then forget about it.

Security briefings should include a section dealing with inadvertent trap doors. Inform the user community that they should advise a specific individual when or if they find such an instance. They should write down what they did so someone else can try and duplicate it.

Focusing on one area and neglecting others. It is easy to gain a false sense of security from behind a locked door, gazing across row after row of encryption devices. It tends to make one feel secure about a frontal attack. Yet in one instance, a bank's entire security was breached by an auditor with a pair of binoculars. Shoulder surfing at a distance, he was able to read the screens of terminals through a window from across the street.

In another case, a hacker was hired to test a major corporation's network security. He was given one week. He broke it in an hour using "social engineering": he talked an innocent user out of his password.

Remembering a little history can be instructive. After World War I, the French built the impregnable Maginot Line to protect themselves against future German attack. When Germany invaded

France in June 1940, Hitler's troops simply went around it to the north. They captured it *from the rear.*

Vendors as unwanted intruders. Occasionally, vendors install a "back door" or a "trap door" entry for maintenance purposes. This sounds reasonable enough, but a back door presents obvious potential problems. First, the way is opened for a hacker. Second, the vendor can modify the software at will without the user's knowledge or permission. In one extreme case, a vendor who claimed nonpayment shut off its software, then in production use, at Avon Cosmetics via a trap door.

Customers have found trap doors by accident. One major corporation licensed a private implementation of a value added network (VAN) for its own use. It turned out that the network provider had included a number of access codes for remote monitoring and maintenance.

The company had not been told of these back doors and had not approved (or disapproved, since the firm did not know about them) their use. The licensee made the VAN vendor remove the access codes and guarantee the firm that there were no more.

The moral: in a contract for services, there should be a clause specifying that there be no trap doors. If the vendor wants a maintenance port, that should be agreed upon in the open. Appropriate security controls should then be installed (e.g., dial-back).

Viruses

Computer viruses are a threat to computers of any type. Their effects can range from the annoying but harmless prank to damaged files and crashed networks. They must be taken seriously.

It would seem that networks are the ideal way to propagate viruses through a system, but the facts are otherwise. The greatest risk is from an infected diskette inserted into a floppy drive.

Viruses have been found in brand-new software diskettes untouched since they were shrink-wrapped at the factory—where they became tainted. For instance, a shipment of Aldus Corporation's Freehand program was contaminated by the Brandow virus. The cause was found to be a Mr. Potato Head game loaded on a production computer after the usual quality assurance checks (presumably including a virus scan) were finished.

Virus scanners and continuously running background programs are no assurance of successful virus detection. Tests show that only a few scanners catch all *known* viruses. Some viruses attack virus scanner programs themselves.

Some newer viruses are self-mutating and have come to be called *polymorphic* (literally, "many forms") viruses. While the telltale signature or bit pattern of a conventional virus is hard to detect, that of a polymorphic virus is harder still because its signature changes with each use.

Viruses with long incubation periods before they inflict damage tend to make us complacent. Subtle viruses, with nonobvious symptoms, compound complacency. An example would be a spreadsheet program that is off a few percent now and then. Equally subtle, the "Dark Avenger from Bulgaria" writes viruses that cause discrete, random bit changes in memory.

All in all, the cost to American industry is over $1 billion a year in direct and indirect costs according to an estimate by the National Computer Security Association. We consider viruses sufficiently important to delegate the next chapter to them. Although the virus threat is hyped wildly beyond its real magnitude, the damage that can be caused in even a few cases can place an entire organization's survival at risk.

Review

- Security is a *people* issue.
- There are many ways by which information may be threatened.
- A security threat may come from outside or inside the organization.
- Of the two origins of threats, the internal threat is greater.
- Commercial espionage, sponsored by a company, has always existed but is now more sophisticated and more highly developed than in the past.
- CI, or competitive intelligence, is distinct from commercial espionage in that CI uses legal ways to collect data.
- Foreign governments represent a new threat on a sophisticated scale and draw on resources unavailable to most private organizations.
- Internal threats include disgruntled employees, terminated employees still on site, and ex-employees.
- System administrators, including LAN and database administrators, are a special risk because their power is usually global and concentrated.
- There are warning signs of pending security trouble in an

employee's behavior. Aberrant behavior is also caused and compounded by organizational problems.

- Unintentional leaks include eyeball leaks, shoulder surfing, and "Maginot Line" thinking.
- Vendors may have back doors or trap doors into their products. These may represent a security risk to your organization.
- Viruses are most often propagated through contaminated diskettes.
- Networks are less often responsible for the spread of viral infections.
- Although viruses infect relatively few networks, they can cause serious damage in those they do. They must be taken seriously.

Notes

1 For further reading, see *The Hacker's Dictionary* or the Jargon file on the Internet or CompuServe. *The Hacker Crackdown,* a book by Bruce Sterling, describes the FBI's effort to prosecute alleged hackers. The book is on-line on the Internet at info.umd.edu in the /inform/computing_resources/netinfo/reading-room/hackercrackdown directory, and is accessed via anonymous.ftp. The text consists of five files for a total of about 600K bytes. To gain insight into how hackers work, consider buying a CD-ROM called *The Hacker Chronicles* from P-80 Systems, Charleston, WV, (304) 744-7322. It consists of an antihacker toolkit and a text-file library. Price: $80.00.

2 Dorothy E. Denning, "Concerning Hackers Who Break into Computer Systems," a speech presented at the 13th National Computer Security Conference, Washington, D.C., October 1–4, 1990.

3 For those interested in further information, see Peter Schweizer's book *Friendly Spies,* Atlantic Monthly Press, 1993.

4 Adapted from G. W. Joseph and Terry J. Engle, "Controlling the Database Administrator," *Information Security,* **3**(2):61, Summer 1993.

Chapter

2

Preventing, Detecting, and Dealing with Viruses

Objectives

At the end of this chapter you will be able to:

- Define a virus and differentiate it from other rogue programs such as worms
- Put the virus threat in its proper perspective relative to other threats
- List the ways viruses propagate
- State the symptoms of the most common viruses
- Discuss antivirus strategies
- Explain antivirus policies and procedures
- Discuss hardware tactics
- Explain the strengths and weaknesses of software antivirus utilities
- State the pros and cons of the several antivirus utility installation methods
- List symptoms that indicate the possibility of a virus infection
- Explain what to do in case of infection
- List common flaws in antivirus strategies
- Understand trends in virus program development and detection

What Is a Virus?

A *virus* is a computer program that copies itself by attaching to other programs, thereby "infecting" them, and performing unwanted actions. Therefore, a virus requires an executable host program to act as a *vector,* a transmission path. A virus replicates and may modify itself every time an infected program is executed. The unwanted actions resulting from a virus can range from a harmless prank at minimum to damage to programs and/or data at the worst. According to National Computer Security Association statistics, the projected cost in lost time, productivity, and file reconstruction in the United States was $2.7 billion in 1994, nearly double the $1.4 billion loss in 1993.

While estimates vary widely, there are about 500 core viruses and about 3,000 variations of them in use around the world. Of these, though, there are about a dozen that are the most prevalent. Virus infections run in waves, with a few viruses infecting many machines over a few months, followed by another series.

Some antivirus software vendors wage a leapfrog war with virus writers, always reacting to the latest virus. Some vendors have now changed their approach, building self-learning or heuristic products that watch for viruslike activity rather than seek out specific virus forms. While this approach leads to some falsely positive results as the program learns one's system, it does relieve network managers of having to update antivirus software every two months or so. We will discuss antivirus software later in this chapter.

There are four main categories of viruses:

File infectors attach themselves to executable files, typically .COM or .EXE files. However, any file containing executable code can become a virus vector.

- *Direct action file infectors* choose one or several other programs to infect each time the program containing it is run. Vienna is a direct action file infector, though most viruses are *resident.*
- *Resident infectors* are placed in memory the first time an infected program is run. They then infect other programs as *they* are run. Jerusalem is an example of a resident virus. Some such viruses are also *fast infectors,* because programs that are simply opened and not even run will become infected. Dark Avenger and Frodo are examples. Ironically, a virus scanner could be the unwitting tool of a fast infector.

- Conversely, a *slow infector* infects files as they are changed or created. Integrity scanners report the change and the user will usually approve it, thinking that the alarm was caused by the modification or creation of the file.
- *Sparse infectors* are subtle. They infect files only occasionally or based on some narrow criteria. Thus they hope to escape detection. Programs that subtly alter the results of calculations also fall into this category.
- *Companion viruses* create a new infected program that looks like the intended program. On exit, the intended program is run to make everything look routine. The giveaway is that the infected file carries a .COM extension where the original carries an .EXE extension (or vice versa).
- *Armored viruses* go to extreme lengths to avoid detection and complicate the reverse engineering needed to build a virus detector.

System or boot record infectors infect executable code found in the system areas of a disk, not considered ordinary files. These are always resident viruses. Examples include Stoned and Michelangelo.

A few viruses are both file and boot record infectors, such as Tequila. You may hear these viruses referred to as *boot-and-file viruses,* or occasionally, *multipartite viruses.*

Cluster or file system viruses corrupt directory table entries in order to load the virus before the desired program.

While viruses can exist in any computing environment, the great majority of viruses are written for the IBM PC and DOS environment, so this chapter focuses on these. Viruses certainly do exist for Macintoshes and to a lesser extent for OS/2 (two million copies in the field). As yet, there are few if any viruses for the newest operating system, Windows NT. The protections described apply to all these environments.

Although not an operating system, MS Windows is virus-resistant (but not virus-proof) compared to a basic DOS environment because most resident viruses are not compatible with Windows' memory management scheme. Often they damage Windows applications upon infection, alerting the user to a malfunction. However, some new viruses *are* Windows-compatible.

Unix-based viruses are quite rare. In part, this is due to the fact that DOS is far more prevalent (80 to 100 million copies worldwide, versus about 1 million for Unix). Virus writers clearly focus on the mass market. Furthermore, the way software is shared in Unix does not lend itself to the spread of viruses. However, a Unix machine running a DOS process such as PC-NFS could be infected by a DOS virus. Conversely, an X86 PC running Unix could be infected by a nonoperating-system-dependent virus such as Michelangelo or Stoned. These are both boot-sector viruses that infect a system at boot time, before the OS is started.

Given the processing power of mainframes, it is no surprise that viruses (written for research purposes only) thrive on a mainframe. However, there are no non-research-created mainframe viruses on record at this writing.

Worms

A virus and a worm are two different kinds of programs. Viruses infect *programs,* while worms infect *memory.* Worms take over memory and deny its use to legitimate programs.

The best known worm incident is that of the infamous Internet worm. In November, 1988, a worm was released into the Internet, the world's largest computer network. Encased inside the Sendmail electronic mail program, the worm took advantage of a weakness in Sendmail's debugging routine to avoid detection. This is called a "back door" attack.

The worm was excessively successful. It impaired the operation of at least 6,000 computers around the world, but was detected quickly because the systems, once infected, slowed down dramatically. Clearly, a major malfunction had occurred, to which systems staff members responded quickly.[1]

Strictly speaking, the term *virus* excludes a *worm* program. However, industry usage has expanded the term *virus* to include worm programs, making the term *virus* synonymous with *rogue programs* or *computer parasites.* We will conform to industry usage but specify worm programs when we wish to exclude viruses from the discussion.

Triggering a Virus

Often, virus programs are not activated immediately after infection. A rogue may be triggered by another program:

- A *time bomb* is a program triggered by a date or time. The Michelangelo virus is a time bomb set to go off on March 6, the artist's birthday.[2]

- A *logic bomb* is a program triggered by some event, perhaps the opening of a file. One disgruntled employee set a logic bomb to go off if his name was removed from the payroll file. Logic bombs, though independent of environment, tend to take place in mainframe-based systems.

Time and logic bombs are planted because:

- The delay makes the program harder to trace.
- The perpetrator can plan the event for maximum effect.
- The delay allows the rogue program to be copied into backup files.

Vendors sometimes implant time and/or logic bombs in their software. If the customer falls behind in payments, or if the customer attempts to copy the program, the bomb is set off and the program or overall system operation halts. The user receives a message telling them to contact the vendor.

Novell uses a logic bomb. NetWare can be copied freely, but if two copies recognize that they have the same serial number, the network stops and the workstations beep. A message indicating duplicate software also appears on each screen.

Trap Doors and Back Doors

Trap doors are intentionally created. A trap door allows access to a system without the usual security controls. Trap doors are often inserted during software development, when security access is a productivity handicap. If all the trap doors are not removed once the system is complete, then there is a possibility that someone will find or use them once the system is in production.

As indicated in Chaper 1, a trap door may be deliberately left in place so that the vendor can provide service and maintenance. At other times it is a means of control, as at Avon Cosmetics where the vendor turned the application off.

Trap doors are intended to be opened only by special command or a set of unique circumstances. These are supposed to be secret.

Unintentionally created access is called a *back door.* The writers of Sendmail presumably did not intend to create a back door into their program. However, many back doors exist. The hacker who invaded the Internet from former West Germany used a back door characteristic of Unix to gain entry.[3]

The Trojan Horse

The usual way a rogue program enters a system is through some innocent-appearing process that happens all the time. One inserts a diskette or performs a system upgrade. The Trojan Horse is embedded in a seemingly benign procedure that could not be performed by a malicious person from the outside, but can be performed by an unknowing person on the inside.

Trojan Horses can carry viruses or worms. They can carry trap doors to be exploited later. A Trojan Horse, then, is not a software program but a ruse designed to penetrate a system so that a program of the penetrator's choosing may become active.

Some examples are as follows:

- Sendmail was used as the Trojan Horse for the Internet worm.
- Two thieves sent a computer tape to NYNEX, the telephone operating company in the northeastern United States, to be mounted on their DEC machine. It arrived in a Federal Express box and looked perfectly official. In fact, it contained a trap door entry program.
- An executable program downloaded from a bulletin board may contain a computer parasite within the program.

Physical Ways Viruses Propagate

Logically, viruses most often propagate as Trojan Horses. Physically, they propagate in the following ways, according to a survey by U.S.A. Research, Inc. (Portland, Oreg.), from the most common to the least:

- Disks from outside the organization (61%)
- Unknown origin (27%)
- Shareware on disks (19%)
- Commercial software (14%)
- Shareware on bulletin boards (13%)
- LANs (6%)
- Electronic mail (3%)
- Wide area networks (WANs) (1%)
- Demonstration disks; repair technicians' diagnostic disks (less than 1%)

An independent study by IBM's Thomas J. Watson Research Center corroborates the U.S.A. Research findings:

- The primary infection path is *contaminated disks,* not computer networks.
- For every 1,000 PCs, a business can expect to contract four viruses a year.
- The average virus infects less than four PCs.

That makes the infection rate nearly 16 per 1,000 or 1.6 per 100 PCs per business in a year.

Recognizing a Rogue Program Infection

What are the clues that indicate a rogue program is present in your system? There are general clues and specific ones. Despite the 2,300 or so viruses in existence, only a few hundred have really spread. Of those, 10 or so are the most common. Accordingly, we present the general symptoms first, followed by the particular manifestations of the most common viruses.

General symptoms

In the U.S.A. Research survey, users were stimulated to use an antivirus screening tool because they observed the first three of these symptoms (in order of frequency from the top down):

- A system error, lockup, or crash
- Screen messages, music, or graphic displays indicating a virus (e.g., Cascade shows letters falling to the bottom of the screen)
- A system slowdown, e.g., longer than normal load times, for no explicable reason
- Too many disk accesses for simple tasks
- Disk access lights come on for no reason
- Unusual system error messages
- Reduced available RAM or disk space for no reason
- Files disappear inexplicably
- Changes in an executable program's file size
- Changes in appearance of screen icons (Macintoshes especially)

Other warning signs were as follows:

- A change of interrupt vectors; reassignment of system resources for no tangible reason.
- Changes in memory map or configuration as the computer boots. Use MEM/C (for DOS) or ResEdit for Macintoshes to see what is normal. If the numbers change for no reason at boot time, this is suspicious.

The study indicates a characteristic typical of many users. They *have* antivirus software but do not *use* it regularly. It is low on the priority list. *After* an infection, when irreversible damage may have occurred, the software is used religiously (at least for a while).

Specific symptoms of the "top twelve" viruses

These symptoms are listed in Table 2.1. Over 90 percent of infected users were struck by a virus on the list. If two or more symptoms listed above or in Table 2.1 occur, then you have special reason for concern.

One of the major problems in fighting rogue programs is that many of their names are similar. This is because they are named informally or by the virus underground. A formal way to classify these viruses is needed so that they cannot be confused.

General Antivirus Strategies

We fight computer parasites in two ways: by having sound policies and procedures in place and by technical means, including antivirus software. *Neither is effective without the other.* Policies and procedures are the first line of defense; technical means are the second.

Policies and procedures

The survey in Table 2.1 gives us strong clues as to what policies we should pursue. Basic antivirus policies are shown in Table 2.2.

Since disks are the main source of infection, antivirus scanning software plays an essential role in keeping parasites out of the workplace. Without a scanning procedure, any continuously running antivirus software is useless.

Consider for a moment what would happen if a virus infected your network or even a standalone workstation. Consider the lost business, the lost productivity, the cost to rebuild lost information (assuming it can be rebuilt at all), and the loss of credibility your

TABLE 2.1 The Top Twelve Viruses*

Virus	Symptoms
Form	Infects DOS system boot records. It contains a bug that can cause data loss, though not intentionally. It will also cause data loss on OS/2 High Performance Filing System hard disks.
Stoned.Standard	Also called Stoned, it was written by a 16-year-old in New Zealand. Intended to cause no damage, it really does by overwriting a sector on hard disks. Variants will damage root or file allocation table directories on 3.5-in disks. The original displayed a message, "Your PC is Stoned!."
Michelangelo	Derived from Stoned.Standard (see above), it infects only 5.25-in disks in the A: drive and the master boot record on C:. It spreads until March 6, Michelangelo's birthday. At that time it overwrites the entire boot drive.
Green Caterpillar.1	Also known as 1575 or 1591, it spreads fast to .COM and .EXE files. The byte count addition may actually exceed 1,575 or 1,591. Activated by DIR or COPY, it may display a green caterpillar.
Jerusalem.Standard	Also known as Jerusalem or Jerusalem-B. It adds 1,813 bytes to .COM files that then infect .EXEs. The original version deletes files on Friday the 13th. Newer versions create various problems on various dates.
Satan	A memory-resident, polymorphic, multilevel (nine possible levels) encrypted virus reportedly composed by a 15-year-old in San Diego, California. Especially well written, it damages .COM, .EXE, .SYS, and .OVE files. Satan also attacks LAN drivers, making LANs inaccessible. Symptoms include growth in COMMAND.COM's file size, typically 3,500 to 5,000 bytes and reduced RAM space by 9K bytes. Date changes, often adding 100 years to the four-digit year in the creation date, and reduced response time are also symptoms. *There are no obvious warnings of infection.* Satan is *designed* to bypass security features built into Novell's NETX.COM workstation shell. Satan will allow only one successful login; after that, the user is locked out. In a NetWare directory, if write permission is denied a user, then Satan cannot infect those files. Unprotected directories, including login files and menus, will be infected. Should Satan attempt to write from memory to a write-protected floppy disk, it will cause a write-protect error message. An unwitting user will turn off write-protect and so allow Satan to corrupt the floppy disk. Satan is especially insidious because it has no obvious symptoms. Without fanfare, it can quietly bring computers and LANs to their knees. To disinfect, every single copy of Satan must be eradicated. Even one copy left "alive" is sufficient to reinfect the entire network.
Cascade.1701	Created in Switzerland in 1987, Cascade.1701 causes text on a color graphics adapter to fall ("cascade") to the bottom line. Cascade.1701 is so named because it adds 1,701 bytes to .COM files.

TABLE 2.1 The Top Twelve Viruses* (Continued)

Virus	Symptoms
Cascade.1704	Adds 1,704 bytes to .COM files. It may cause random reboots. However, it does not drop letters to the bottom of a screen.
Dark Avenger.1	Adds 1,800 bytes to .COM and .EXE files, but causes no damage.
Dark Avenger .4	Adds 2,100 bytes to .COM files. It is also known as Eddie II, V2100, and 2100.
Frodo	Adds 100 years to the DOS date and time stamp. It infects .COM and .EXE files quickly after the file is opened. Frodo is also known as 4096 because it adds 4,096 bytes to the file size. It is sometimes called 100 Years and Stealth, the latter because once in memory, the increased file size does not show.
Yankee Doodle 2885	Also known as TP44VIR, it is the 44th version of a virus written by "TP" of Bulgaria in September, 1989. It plays "Yankee Doodle" at 5 P.M., but causes no damage (except to your nerves).

*Also see the Notes at the end of the chapter.

organization might suffer. Weigh that against the cost of doing some fairly simple and inexpensive things to give your organization basic coverage. The key is to present the entire matter to management as a *business* case, not as a *technical* case.

Consider that a study in the United Kingdom by Price Waterhouse found that a business taken off line for *three days* stands a *60 percent chance* of going out of business. After a month, the figure rises to *90 percent*.

A common fallacy is to look at a server as just another PC. Say a server is worth $5,000. What this view neglects is that hundreds of thousands of dollars' worth of information may be buried in it. Thousands of hours of work, virtually irreplaceable, may reside within the server and individual workstations, whether the latter are networked or not.

Specific Antivirus Tactics

In a survey of 622 companies of different sizes and in different industries, U.S.A. Research found the following virus prevention strategies in use:

TABLE 2.2 Antivirus Policies and Procedures

✓ Build any system from original, clean master copies. Boot only from original diskettes whose write protection has always been in place. Use only original diskettes until virus scanning is performed.

✓ *Allow no disk to be used* until it has been scanned on a standalone machine that is used for no other purpose and not connected to a network. Update the scanning software frequently. Some users use several scanning tools in sequence executed by a batch file.

✓ Write-protect all diskettes with .EXE or .COM extensions. (Note, however, that what makes a file executable is its *format,* not its *extension*).

✓ Have vendors run demonstrations on *their* machines, not *yours.*

✓ Shareware on disk or downloaded from a bulletin board is one of the most common infection sources. *Try not to use shareware at all.* If you must, scan thoroughly before use.

✓ Commercial software is occasionally a Trojan Horse for viruses or worms. Scan before any new software is installed.

✓ Insist that field technicians scan their disks on one of your machines before they use any of their disks in your system.

✓ Use workstation and server antivirus software.

✓ If possible, place all executable files on a separate, read-only disk. Software write protection *can* be bypassed by some viruses.

✓ Using GUARD or an equivalent program, create a special master boot record that makes the hard disk inaccessible when booting from a diskette. This ensures that the hard disk cannot be contaminated by the floppy disk, since the CONFIG.SYS and AUTOEXEC.BAT files containing scanning commands have now been bypassed.

✓ Consider encrypting files and then decrypt them before execution.

✓ Be sure that bridge, router, and gateway updates are authentic. They are ideal places for Trojan Horses to hide.

✓ *Backups* are a vital element of antivirus strategy. Be sure that you have a sound and effective backup plan in place.

✓ *Obtain strong management support.* Most managements strongly endorse antivirus policies, but many fail to back them up with budget.

✓ *Educate* users so they will heed these policies and procedures. They apply to management, too.

✓ *Publicize* virus contaminations. The message: it *does* happen.

✓ Make these policies and procedures *accountable* during employee performance reviews.

✓ Review your antivirus policies and procedures at least once a year.

- Antivirus software (63%)
- Computer access policies (41%)
- Off-network computers for outside communications (20%)
- No strategy (20%)
- Write-protected floppy tabs (19%)
- Other utility software (9%)
- File encryption (4%)
- Other (3%)
- Antivirus hardware (1%)
- Strategy unknown (1%)

Hardware tactics

Four hardware tactics reduce the risk of infection.

1. *Use workstations without floppy disks.* Since this is the major way workstations become infected, eliminate them altogether when you can. This is not possible in all applications, but it is in some where all necessary software is on the server. If a disk needs to be loaded, let it go through the network administrator, who scans it for parasites first.

2. *Use remote booting.* There is a place for a local or remote boot programmable read-only memory (PROM) in most LAN network interface cards. With remote booting from a master boot file on the server, there is no need to boot locally, especially from a floppy disk that might have a virus in its boot sector. A secondary advantage is that if the system needs reconfiguring, it can be done from the server boot file instead of having to change each workstation's configuration individually by hand.

3. *Use a hardware-based password.* As an example, Centel Federal Systems offers hardware that prevents booting without first entering a firmware-based password. This is a security tool to prevent deliberate tampering or unauthorized access.

4. *Use write-protect tabs.* They will stop a virus from infecting a floppy disk, provided that:

- The disk was not previously infected
- The tab was not turned off and on again, allowing a virus in
- The write-protection mechanism is not defective

Note that a write-protect tab will not stop a virus that is already running.

Software tools

Software tools are by far the most common antivirus tool. They have had a significant impact on smothering the virus outbreaks of a few years ago. They should be used *proactively* and not *reactively.*

All software tools encourage a false sense of security. None are perfect. Unless updated periodically (about every two months), some will not detect the latest virus strains. Others detect viruses where there are none, causing needless disruption. Dr. Fred Cohen, who coined the term *virus,* has shown in theory that every virus detector must have false negatives, false positives, or both. However, his proof does allow for no false negatives and just a few false positives. The reality is that false positives are rare since good virus writers pick strings that do not match common programs. False negatives are more common, usually due to a new viral signature.

If a scanner detects a virus, it is important to verify it with another scanner. Using the wrong technique to correct a virus can cause irretrievably lost data.

It is important to know what the software can and cannot do. Only then can you avoid the ever-insidious "Maginot Line thinking." Most software tools are designed to run in workstations, but server-based software tools are now coming on the market.

Server-based software is the last line of defense. Do not forget that dial-in users and temporary connections to the LAN via visiting notebook users may not have any workstation protection at all. Server-based software thereby takes on added significance.

Types of Antivirus Software[4]

As Fig. 2.1 shows, there are three basic forms of antivirus software: scanners, active monitors, and integrity checkers. They may be executed as either device drivers or terminate-and-stay resident programs.

Scanners

Scanners look for sequences of bits called *signatures* (see Fig. 2.2) that are typical of virus programs. Scanners examine memory, disk boot sectors, executable, and command files for bit patterns that match a *known virus.*

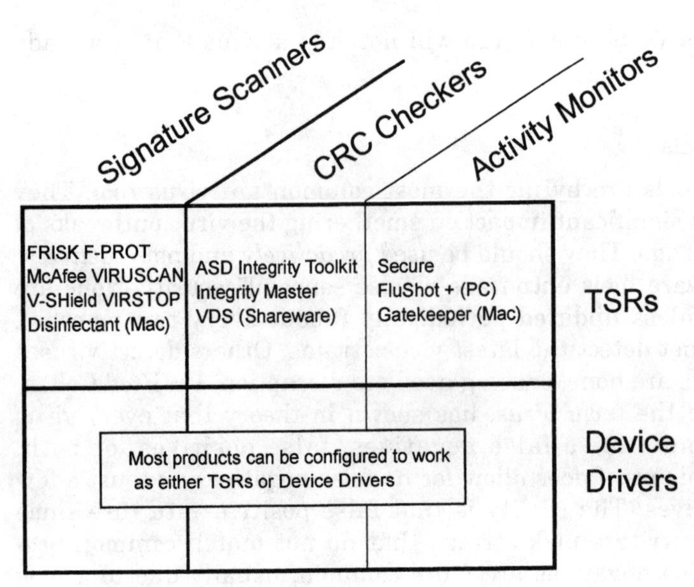

Figure 2.1 Types of antivirus software.

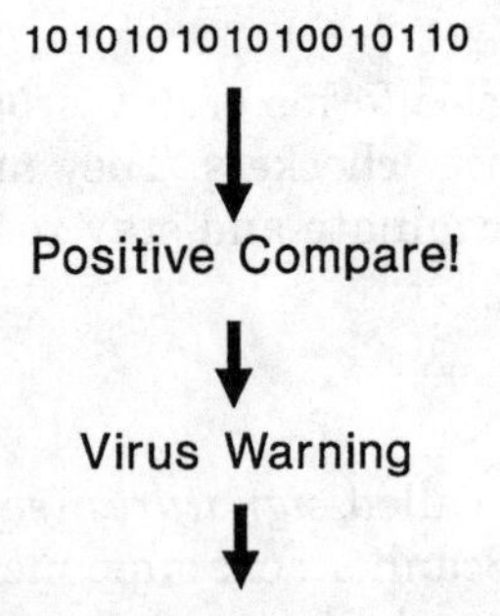

Figure 2.2 Signature scanning.

Most current antivirus software is based on scanner technology. Scanners therefore need to be updated periodically to remain effective. Many vendors have an update subscription service.

New viruses can fool outdated scanners. *Polymorphic* (literally, "many forms") viruses have no unique signature: they are encrypted. Polymorphic viruses contain a self-garbling mechanism that generates random decryptor loaders. These loaders contain a randomly ordered set of instructions. Polymorphic viruses consequently change every time they are executed. Only by scanning a file a byte at a time can such a virus be caught.

Some viruses, especially those written by the Bulgarian "Dark Avenger," use a mutation generator to change their signature. In response, newer scanners look for signs of the mutation engine, too.

Scanners must be run frequently to be effective. Once run, they will not run again until the machine is rebooted or the program is run manually. Scanners are usually run from AUTOEXEC.BAT in workstations. However, it may take too much time to run a scan on a whole hard disk every time it boots. For that matter, the scanner will be ineffective if the command is in AUTOEXEC.BAT but the machine is rarely booted.

Users quickly realize that time needed for scanning during boot-up is significant. The impatient ones may find that hitting the escape key may circumvent the scan and speed boot-up.

Scanners that run as Terminate and Stay Resident (TSR) or device driver programs do not have this startup problem. Instead, they scan programs automatically as they begin to run. If the scanner is a TSR program, there may (of course) be memory usage problems and program conflicts. We will discuss the pros and cons of one-time versus TSR or device driver software shortly. Scanner programs include McAfee's Scan and Symantec Norton's Anti-Virus.

Active monitors

Active monitors may also be TSRs or device drivers. As Fig. 2.3 shows, they intercept DOS and ROM basic input–output system (BIOS) calls, looking for viruslike actions (such as asking for disk formats).

Active monitors can be annoying because they cannot distinguish between a user request and a program (or virus) request. As a result, users are asked over and over to confirm things like formatting a disk or deleting a file or set of files.

Newer versions can be tuned to eliminate most false positives. For a more direct approach than signature detection, it seems

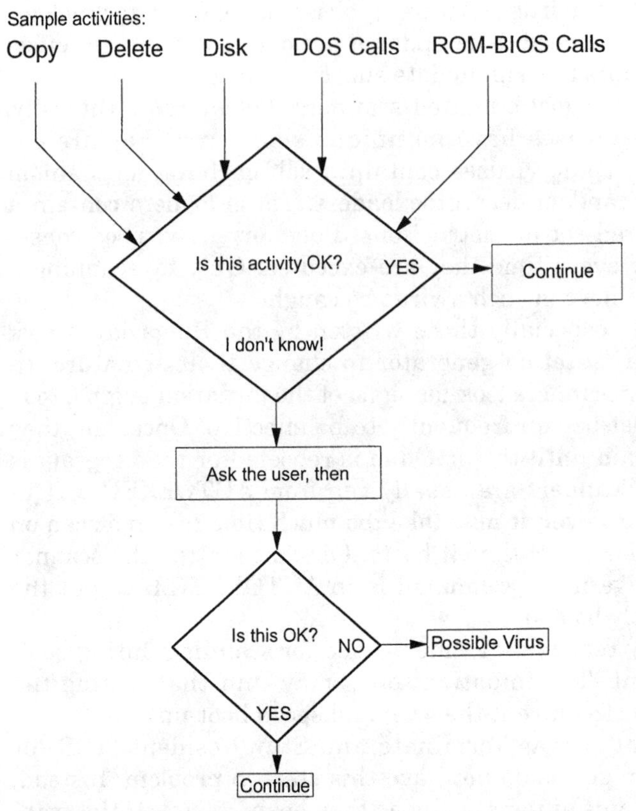

Figure 2.3 Activity monitoring.

preferable to detect viruses by what they do rather than by what they look like (assuming that one knows what they look like in advance). Examples of active monitors include Virex/PC and Flu Shot+.

Integrity checkers

As shown in Fig. 2.4, integrity checkers compute a binary number on a known virus-free program that is then stored in a database file. The number is called a *cyclical redundancy check* or CRC. Sometimes a cryptographic "hash value" is calculated instead, but its function is the same as a CRC.

When that program is called to execute, the checker computes the CRC on the program about to be executed and compares it to the number in the database. A match means no infection; a mis-

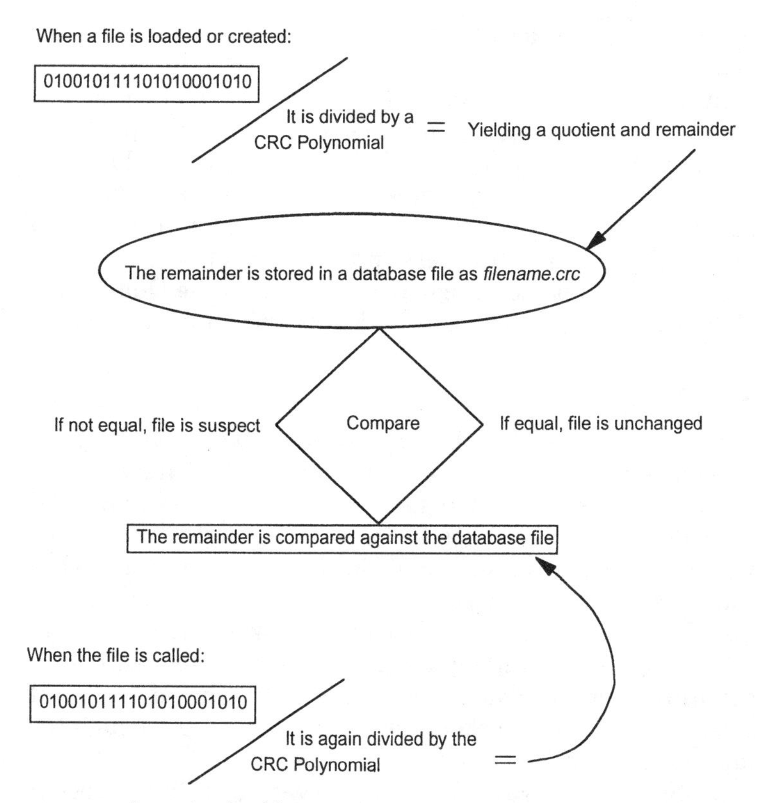

Figure 2.4 Integrity checking.

match means that a change in the program has occurred. Sometimes the number is attached to the end of the executable program.

A change in the program could mean a virus within it. If the program was legitimately changed, the CRC checker cannot know. Integrity checkers take advantage of the fact that executable programs and boot sectors do not change very often if at all. File integrity checking is especially useful in Unix systems, where unauthorized file changes (though not usually due to viruses) are common. A program called COPS is available free from cert.org via anonymous file transfer protocol (FTP) to perform this check as well as others.

There is no need to update integrity checkers since they do not look for signatures. They are particularly good at spotting *stealth* viruses. Stealth viruses record the original file size and alter the file allocation table to show the pre-infection size. This defeats antivirus programs that look for file size changes as indicative of a viral infection.

Even an integrity checker will not catch a memory-resident virus. Such viruses are invisible to almost every antivirus tool.

A problem with integrity checkers is that they give no clue as to what if any virus is in the program. If the program is not supposed to change and it does, then it should be deleted and a backup installed, just to be sure. The question remains as to *why* the program changed. Examples of integrity checkers include Quarantine (On Disk Software, N.Y.) and Integrity Master (Tallahassee, Fla.).

Combined methods

Some products include all three methods, allowing the user to choose which is best. Microcom's (Norwood, Mass.) Virex/PC and Central Point's (Beaverton, Oreg.) Anti-Virus are examples. All three methods can give false positives, meaning that the antivirus detectors think a virus exists when there really is none. Such a false positive is caused by a virus signature that occurs normally in a program. The signatures are very short, allowing much opportunity for "aliasing," or mistaking a legitimate binary sequence for a virus signature. Similarly, active monitors wait even after honest operations, and integrity checkers could ring a bell even though the program change is legitimate.

Virus protection software for LAN servers is a fairly recent development. In servers, a continuous scanner and an integrity checker are the best tools. Intel's LAN Protect does part of the task as a continuous scanner, but for NetWare 3.1X only. We will come back to the topic of server virus protection.

TSRs versus device drivers

Continuous scanners are implemented both as device drivers and TSRs. Each approach has its advantages and disadvantages, as shown in Table 2.3. The major differences are in the degree of virus protection they provide and in the convenience they offer.

Scanner device drivers are loaded in CONFIG.SYS, usually as the first line. This prevents a virus from entering during the boot process, as CONFIG.SYS runs before AUTOEXEC.BAT. Device drivers must exist individually on each workstation.

TSRs are loaded in AUTOEXEC.BAT. However, since the boot process has already occurred, a virus could slip in before the TSR is loaded. One advantage of TSRs is that they may be downloaded from the server.

TABLE 2.3 Antivirus Scanners: Device Driver or TSR?

Device driver	TSR
Loads in CONFIG.SYS (first line)	Loads in AUTOEXEC.BAT
No window opening for a virus	Leaves an opening for a virus
PC-resident	PC- or server-resident (sharable)
Cannot be installed via login script	Can be installed in the workstation from the server via a login script
Protection whether logged in or not	No protection unless logged in
Cannot be removed without CONFIG.SYS change and reboot	Can be removed from memory without change to AUTOEXEC.BAT or reboot

Windows applications should call an antivirus program in the first line of AUTOEXEC.BAT. Then Windows will copy the virus scanner along with DOS into each "virtual machine" window that it creates.

Both types of antivirus programs are memory-resident, so small size is desirable. Both should scan programs during loading and execution. Copying and writing should also be monitored.

Any virus-scanning software should prevent infected files from running. Some scanners go further and automatically delete any virus found. In the interest of security, only the administrator should be able to reconfigure the scanner configuration via password. Be aware that these products provide mainly file protection. A floppy disk inserted in the machine may contain a boot-sector virus that will not be caught.

Neither program type above is intended to run on a server, but some scanners provide *server protection,* as Fig. 2.5 shows. That is, they monitor uploads to the server and downloads from it, looking for viruses. If they spot one, a message appears and the virus is not loaded. Some NICs, as from Intel, also scan traffic.

As with other antivirus programs, spending more money does not necessarily mean getting a better program. Read all available information, such as Ziff-Davis's *Computer Select,*[5] and especially *several* trade press reviews before buying.

Server-based antivirus software

In principle, scanning all the workstations prevents infecting the server. However, workstation scanning is insufficient by itself because of remote access through bridge–router connections as well as dial-in access and direct wire access via visiting notebook users.

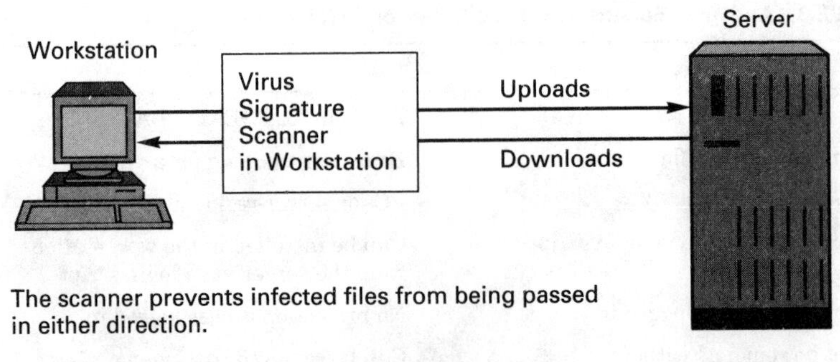

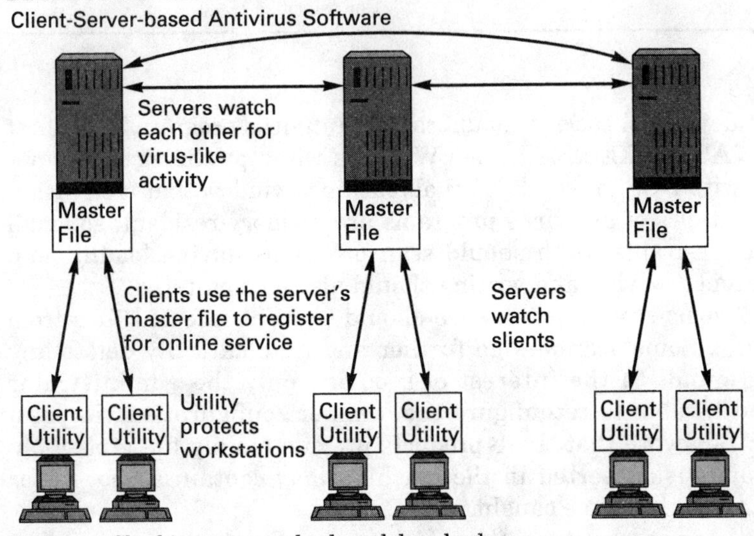

Figure 2.5 Checking server uploads and downloads.

Scanning a server as one would scan a workstation must usually be done during idle time because such scanning is very time-consuming and degrades network performance. Usually a scan can be scheduled to run off-hours.

For NetWare 3.X and 4.X users, a variety of NetWare Loadable Modules (NLMs) and Value Added Processes (VAPs) for NetWare 2.X are available.[6] As plug-in software options, they monitor server operations continuously.

Some NLMs have client–server capability where servers watch one another as well as all clients. Servers keep master file information that they share with clients. A client-side utility may even need to be running before a login can occur. The utility protects the client machine from viruses and uses the server's master file data for "registration" on line. Temporary connections via notebook machines are also watched by the server, but the machine need not necessarily be a client-state machine.

In various tests, antivirus NLMs have not gained sterling reviews. Some NLMs failed to detect certain viruses, adversely affected performance, or caused lockups. In some cases, the NLM caused more problems than a virus might have.

If considering an NLM or equivalent, evaluate these factors:

- Ability to detect viruses
- Possible adverse impact on network performance
- External package security and internal self-check ability
- Reporting and alert mechanisms (e.g., to printers, screen warnings, pagers)
- Ease or difficulty of use
- Vendor support, especially updates

Also see Table 2.4.

Software summary

Using our perimeter approach on a small scale, do the following:

- Select a ROM BIOS password and use only a known good boot disk.
- Employ software that detects boot-sector viruses and changes in boot sectors of any kind.

TABLE 2.4 Antivirus Software Checklist

✓ Buy a multifunction package that includes scanning, CRC checking, and virus removal. All are useful.
✓ The more viruses detected, the better the package.
✓ The software should identify both common and rare viruses in files and boot sectors.
✓ The software should find master boot-sector viruses.
✓ An easy-to-use display format is very important. It should lead you through a proper course of action in case a virus strikes.
✓ Information about what the virus is and does should be available.
✓ Scanners should run quickly. Some products will do a quick scan and a full scan as well.
✓ The software should not report false alarms.
✓ The software should run a self-check to make sure it has not been compromised.
✓ Be aware that there can be substantial performance differences between packages.
✓ Consider all aspects of a package, not just a few.

- Use at least two scanning tools, one at boot time and one running in the background. Keep signature scanners up to date.
- Where possible, use software and hardware locks to prevent writing to disks, whether fixed or removable.
- Setting non-networked DOS files to "read only" is a good idea, but many viruses will simply override it.

LANs as Virus Vectors

Properly established LANs make poor propagation paths for viruses for the following reasons:

- LANs are set up to have access to a shared set of read-only executables. This minimizes diskette sharing, the prime viral vector.
- LANs give minimal privileges to each user.
- They minimize files written by one user and readable by others.
- LANs use login scripts to authenticate workstations before network access is allowed.
- Password protection is of no value in virus protection. Privileges accorded your password are acquired by the virus.

Conversely, poorly setup LANs can propagate a virus throughout the network.

NetWare virus resistance

Novell NetWare-based LANs have some virus resistance because the login process and virus propagation interfere with each other to some extent. One source of interference is *interrupts*. Interrupts ask the central processing unit (CPU) for attention. Depending on the importance of what the CPU is doing, the CPU may put what it is doing aside and service the interrupt if it is important enough.

Accordingly, there are *high-level* interrupts and *low-level* interrupts. Low-level interrupts usually come from hardware and before COMMAND.COM is loaded. Boot-sector viruses usually use these. Conversely, software uses high-level interrupts after COMMAND.COM is loaded. File viruses tend to use high-level interrupts.

In NetWare, IPX and NETX use these same high-level interrupts. One result may be a lockup because the two are fighting to get the interrupt. If the virus is using the interrupt when IPX and

NETX are loaded, they may not load and no network connection will result. IPX and NETX may also override the virus and gain the interrupt, preventing the virus and connecting successfully to the network.

Boot-sector viruses

Boot-sector viruses propagate by copying from infected diskettes into a workstation. Another propagation method is to compress a file or diskette; this shields the virus from a scanner. If copied onto the network, it can go anywhere. Infection by floppy disk exchange is illustrated in Fig. 2.6.

Since boot-sector viruses use low-level interrupts, NETX and IPX will load normally. However, since NetWare's file structure is not DOS, a boot-sector virus cannot infect the server's drive.

The only way a server drive can be destroyed is if the server has a DOS partition. If an infected boot-sector floppy is used to boot the server, then it can get onto the DOS partition and infect its boot sector.

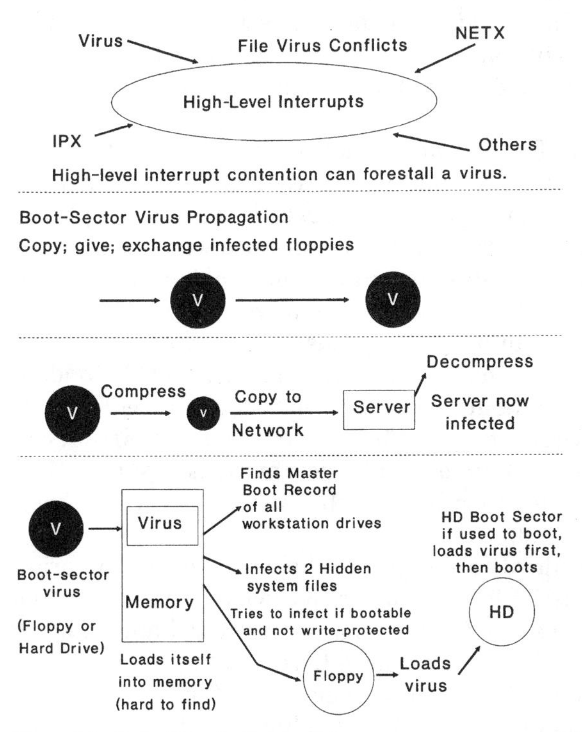

Figure 2.6 Infection via floppy disk.

TABLE 2.5 Virus Infection Checklist

✓ Stay cool.
✓ Cease operation.
✓ Start a log and take notes as you go through the following steps.
✓ Disconnect the infected machine from a LAN; remove diskettes and mark them.
✓ Observe what the machine does. Try to identify the virus.
✓ Run the antivirus scanner. Use an external write-protected disk.
✓ Copy infected files to a diskette. Delete infected files.
✓ Clean with a virus scrubber or restore from backups.
✓ Scan everything that might have carried the infection.
✓ Look for the source.
✓ Review and revise your procedures.

The bottom line: while networks have some protection, it is not good enough to trust.

What to Do in Case of Infection

The first thing to do is to take a deep breath and *stay cool*. People usually overreact and disconnect entire systems. This causes unnecessary down time. A virus usually affects executable code, not data. You can safely wait until the working day is over to clear the problem. See Table 2.5 for a short list of what to do and what not to do.

Cease all operation. Execute no further programs in the hope the virus will not spread farther than it already has. Do not turn the machine off; turning it on again may cause the virus to replicate again. Put a note on the screen so that no one else will use it.

Now disconnect the offending machine from a network if it is on one. Remove any disks. It may be too late anyway, but do not give the virus any additional propagation paths.

Once isolated, observe what is going on. You may be able to identify the virus. It may announce itself. ("Your PC is Stoned!" seems pretty clear.) Find where the virus has copied itself. See if you can regain control of the machine and run the antivirus scanner. Most scanners have a self-detection feature that indicates if they themselves have been infected.

Copy the infected file(s) to a separate diskette. Delete the original file(s). The deleted files should be overwritten with random ones and zeros. This way, the file cannot be undeleted and so bring the virus back to life. You have a choice here. You can restore from a known clean backup or try to scrub or clean the file.

Some antivirus tools have "scrubber" software that is supposed to eradicate a virus. Some are better than others. Many experts distrust virus scrubbers because none restores the exact state of the file before infection. This includes the original creation date and time, file size, and all headers. True recreation is sometimes impossible since bugs in viruses commonly eradicate such information irretrievably.

A file not located on the same clusters as before infection may not run because the cluster numbers are used sometimes as a self-checking or copy protection device. Some sparse infectors such as Phoenix infect very subtly, making it hard to tell which, if any, files have been corrupted. Given the choice, restore from a known good backup.

If the infected file contains data that cannot be restored or rebuilt, then using a scrubber may be the more pragmatic if not necessarily the technically better choice. Should the scrubber leave the file with more bytes than before it was infected, that is usually not a problem. If it runs all right, use it. However, if there are *fewer* bytes, restore from a backup.

Obviously, scan every machine that could have been contaminated. Remember the notebook that was temporarily connected last week?

Boot-sector infectors can be bypassed by using a known clean system diskette, but it is wise to check all diskettes that were in the machine for infection. Replace the master boot record by using FDISK/MBR followed by executing the SYS command to replace the DOS boot sector. Alternatively, and perhaps more simply, use a backup.

It is important to remember that inserting a nonbootable disk at boot time still causes the master boot record to be read. Thus a virus can enter even if the disk is a nonsystem disk. Any formatted diskette is a risk.

Try to find the source of the infection, so it cannot happen again. Scan everything you can (this could be quite a task). Sometimes a worker has brought in a disk from home that caused the infection. If so, backups prior to that date are safe. When the dust has settled, review your procedures and make necessary changes.

Common Flaws in Antivirus Strategies

The biggest and most obvious flaw is that there may be no strategy at all. With no antivirus programs, scanners, or tools of any sort, you are defenseless. There being no roadblocks, the virus can run its course.

Having a good backup and disaster recovery scheme is essential. Virus scrubbers cannot repair lost data. As you saw, the quality of scrubbed programs remains suspect. Even if the virus is eradicated, the program may not run properly after scrubbing.

When running backups, consider the following:

- Do not back up executable code; rely instead on the original diskettes.
- Do, of course, back up data, text files, and source code. As yet, no source code viruses exist.
- Save suspect software (i.e., shareware) in non-executable form, such as in .ZIP format. Back it up only once.
- Archive original commercial software, then save it on the medium it came on. Again, back it up only once.
- Do not back up reconfigured executable code; reconfigure it later if necessary. This does not apply to new versions of executable code, which should of course be backed up.

Procedures tend to get sloppy with time. People stop scanning diskettes from the outside. When nothing happens, as is usually the case, complacency is encouraged and people become still more lax. It is important to keep antivirus procedures in the forebrain of every user. Publicize it on login screens, in company newsletters, and on performance reviews. It is a never-ending task. Administrators and MIS people must set the example. If they do not observe the rules, no one else will, either. This is especially true for management.

The standing rule of network security applies:

Never let your guard down.

Trends In Virus Program Development and Detection

Virus program development

Some rough calculations show that between two and six new viruses are written every day. As far as is known, there are between 2,000 and 2,300 viruses in existence, but less than 200 are common. Many are variants of old originals such as Jerusalem B and Stoned, rather than all-new ones. Many originate in Eastern Europe, especially Bulgaria, and Russia. At the other end of the spectrum, simple possession of a computer virus in mainland China is a crime.

Virus writers are all too innovative. Viruses now attack antivirus programs. The mutation engine and the nuclear encryption device,

(NED) eliminate the signatures that scanners look for. The signature scanner alone is no longer enough.

A virus development tool, the Virus Construction Laboratory, or VCL, is now on bulletin boards around the world. Written by the "Dark Avenger," it allows almost anyone to create a virus. A newer VCL version uses self-encryption to hide any possible signature. By one estimate, some 1,500 copies of VCL have been downloaded from bulletin boards.

Antivirus software development

The common and long-term key seems to be software that looks for misbehavior. Activity monitors detect viruslike activity and report it. Newer systems, such as Trend Micro Devices's (Los Angeles, Calif.) PCRx, can be tuned to minimize false alarms.

Assessing the risk

An interesting contradiction is that while the number and sophistication of viruses are increasing, the actual number of infections is declining from a peak in 1990–91. The decrease is attributed to aggressive virus awareness campaigns.

In the words of U.S.A. Research: "These dramatic and future decreases in virus infections or reinfections are *highly* [emphasis theirs] dependent on the continued aggressive trends in antivirus education...and antivirus measures used by computer users and computer managers. Without a strong antivirus response the current and future number of computer virus infections or reinfections would be much higher."

Fortunately, some of the newer viruses are relatively benign. Newer viruses are often more annoying than destructive. However, damage does occur and it is often not possible to limit the damage to a small area should you be struck by a malignant virus.

Loss experience

What about the damage? A 1993 online survey by the Computer Security Bulletin Board System[7] found that 64 percent of the respondents had been attacked by a virus in the last year. Half were barely affected, but 6 percent sustained losses of over $100,000 and took more than three days to recover. Some 28 percent of the infections cost under $100,000.

The largest virus attack reported in the survey was at a financial institution. It infected 34 networks and took 45 working days to recover. The company lost over $300,000 and never found the source.

The bottom line

Although viruses are ever more sophisticated, so are the tools to catch them. For the moment the antivirus tools have the upper hand. This is not the time to become lax. Attacks will not go away. Those who are hit can sustain losses as high as $100,000 or more. Exposure to viruses is simply not an affordable risk.

The migration of systems to LANs and client–server computing will create more systems subject to attack. Mission-critical applications will run on these systems. Exposure of any kind to viruses could mean severe losses.

All Viruses Are Not Bad

Viruses can fight. During the Gulf War, the National Security Agency reportedly prepared a virus that crippled the Iraqi air defense computer system. The virus cleared information from computer screens as soon as it was called up.

U.S. intelligence agents reportedly intercepted a French-made printer and replaced an original chip with a virus-contaminated one. Since the contamination was in a printer and not in the system itself, it apparently bypassed Iraqi computer security routines.

Dr. Fred Cohen, who coined the term *computer virus* in 1983, wrote[8] that "the features that make computer viruses a serious threat to computer integrity can also make them a powerful mechanism for the reliable and efficient distribution of computer resources."

He goes on to illustrate: international computer researchers use the Lenstra–Manasse algorithm to factor large numbers. Asking permission first via electronic-mail (e-mail), the viruslike algorithm uses otherwise unoccupied resources. Once results are obtained in a machine, the algorithm reports its results back to a central site via e-mail. The central machine then distributes newer and matured parts of the problem. This method was used to break the Rivest–Shamir–Adelman public-key algorithm by a team led by Bell Communications Research. (See Chapter 3.)

The system is efficient, using otherwise unused resources. It uses parallel processing. It is reliable, because there is no single failure point. Mostly, though, it is *legal,* because it *asked* first.

Dr. Cohen used a virus in a law office bill collection system. Each time there is a new case, it spawns a bill collector virus, designed to collect one bill. When the bill is paid, the virus is killed.

Summary

Viruses exemplify "Maginot Line thinking." Companies install antivirus software and then neglect the rest of the security picture, especially the inside threats that are far more common.

Of course, this does not mean that viruses should be neglected. They are part of an overall security strategy. Those affected by a virus have lost between $10,000 and $300,000 in direct and indirect losses. No organization can afford this kind of exposure when the ounce of prevention is comparatively inexpensive.

A complete antivirus policy should include:

- *Preventing* infection of your software
- *Detecting* a virus if it gets past the previous step
- *Controlling* contamination (e.g, by limiting permissions)
- *Recovering* from an infection by using scrubbers or backups

As you can see, this is a *perimeter* approach, starting from the outside and working in. It is a good approach not just for virus protection, but for security in general as was illustrated in Chapter 1.

Review

- Viruses copy themselves by hitching a ride on another executable program.
- The actions of a virus can range from harmless to deadly but are always unwanted.
- Viruses and worms (that infect memory) are often propagated by a Trojan Horse attack, a seemingly innocent action containing hidden consequences.
- Viruses propagate more often via disk than any other medium, including networks. *No formatted diskette is safe.*
- Infections may announce themselves. Sometimes the system crashes or slows down for no obvious reason. Some viruses have subtle effects.
- Most infections are caused by a few well-known viruses.

- An effective antivirus strategy uses several tactics, not just one.
- The tactics are prevention, detection, decontamination, and recovery.
- Antivirus software scans for signatures, watches for improper activity, or checks files for possible contamination. Use at least two, especially the latter two.
- Antivirus software may be implemented in workstations as either TSRs or device drivers. Each has its own advantages and disadvantages.
- Server-based software is highly desirable, especially since all workstations may not have antivirus software, particularly traveling notebook or laptop machines.
- If a virus strikes, stay cool. Follow the checklist given earlier.
- Having backups and a disaster recovery plan is essential antivirus strategy.
- Antivirus education and software have reduced virus incidents. However, losses as high as $300,000 are known to have occurred.
- Increasing reliance on client-server systems will increase our exposure.
- Not all viruses are bad—but most are.
- *Never, never, let your guard down.*

Notes

1 For a description of what happened, see E. Spafford's "The Internet Worm Program: An Analysis," *Computer Communication Review*, vol. 19, no. 1 (1989), Association of Computing Machinery SIGCOM. Also see *A Tour of the Worm,* Proceedings of Winter 1989 USENIX Conference, Usenix Association, San Diego, CA, February 1989.

2 This is a perfect case of mass hysteria triggered by the media. For the vast part, the Michelangelo time bomb was a fizzle. In fact, some LAN users who purchased multiple scanners caused their own conflicts. These conflicts were nearly as disruptive as the virus they did not have in the first place. It is worth noting that after March 6, most installations returned to business as usual. Ironically, Michelangelo's succeeding birthdays have been all but ignored.

3 See Clifford Stoll's book, *The Cuckoo's Egg,* Doubleday, Garden City, N.Y., 1989.

4 A list and evaluation of many antivirus software programs can be obtained by anonymous FTP from cert.org in pub/virus-l/docs/reviews. The file name is slade.quickref.rvw. The same site (in the /docs directory) offers National Institute of Standards and Technology (NIST) Special Publication No. 500-166, "Computer Viruses and Related Threats: A Management Guide," in both text and PostScript versions. NIST offers a free document available via dial-up entitled "A Guide to the Selection of Anti-Virus Tools and Techniques." Call (301) 948-5717 to obtain a copy from their online system. Freeware and shareware is available from several sites, again by anonymous FTP: DOS programs can be obtained from WSMR-SIMTEL20.ARMY.MIL Path: PD1:<MSDOS.TROJAN-PRO> and Macintosh antivirus programs can be found in PD3:<MACINTOSH. VIRUS>

5 Ziff-Davis, *Computer Select.*

6 The following is a list of some antivirus NLM vendors:

Brightwork Development	SiteLock (NLM and VAP)
Central Point Software, Inc.	Anti-Virus
Cheyenne Software, Inc.	InocuLAN (NLM and VAP)
Intel PCED	LANDesk Virus Protect
Intel	LANProtect (Client–server; MAC)
McAfee Associates, Inc.	NetShield
Ontrack Computer Systems, Inc.	Anti-Virus Toolkit

A non-NLM package for servers is *Symantec Norton Antivirus* (Cupertino, Calif.), which works with StarLAN, Banyan VINES, NetWare, IBM LAN Server, and Microsoft LAN Manager.

7 The Comsec Bulletin Board is at (415) 495-4642. Creative Strategies, Inc. [(408) 748-3400] analyzes results from bulletin board system surveys.

Further virus information sources are as follows:

National Computer Security Association, Carlisle, Penn., publishes the "Computer Viruses" report. It is reasonably accurate and up to date.

International Computer Security Association, Washington, D.C.

The VSUM file describes DOS viruses and can be downloaded from most Internet Archive sites except SIHTEL20. While most known viruses are listed, its accuracy is sometimes questionable. Sole reliance on VSUM is not a good idea.

The Computer Virus Catalog (CVC) is published by the Virus Test Center in Hamburg, Germany. DOS, Macintosh, and Unix viruses are covered, though the DOS section is far less complete than VSUM. Use anonymous FTP to download a copy through cert.org in the pub/virus-l/docs/vtc directory.

The Computer Emergency Response Team (CERT) Coordination Center at Carnegie-Mellon University in Pittsburgh, Penn., is available if you experience a major infection. Some vendors also offer rapid-response teams. CERT can be reached via the Internet at cert@cert.org or via telephone 24 hours a day at (412) 268-7090. Look in the pub directory for an array of virus- and security-related papers and references, especially the frequently asked questions (FAQ) files. The following is an edited excerpt from the FIRST-CO. file at cert.org:

> The CERT Coordination Center is a founding member of the Forum of Incident Response and Security Teams (FIRST). The following is a list of other incident response teams and their constituents.
>
> This is a list of contact information for incident response teams participating in FIRST. This list is updated periodically; a master copy of this list is available from the FIRST Secretariat via anonymous FTP at csrc.ncsl.nist.gov (129.6.54.11), file pub/first/first-contacts, or by sending e-mail to docserver@first.org with the message: send first-contacts.
>
> If you cannot figure out who to call, contact a response team or the FIRST Secretariat at (301) 975-5200 or first-sec@first.org.
>
> The following list is organized by constituency. "24/7" indicates that the phone is monitored 24 hours, 7 days a week. E-mail addresses are reachable via the Internet. All telephone numbers are preceded with the appropriate country code; U.S. callers must add "011" to call non-U.S. and non-Canada numbers.

Constituency: Internet
Response team: CERT Coordination Center
E-mail: cert@cert.org
Telephone: +1-412-268-7090, 24/7
FAX: +1-412-268-6989

Constituency: Air Force—U.S. Air Force
Response team: AFCERT
E-mail: Imreyes@dockmaster.ncsc.mil
Telephone: +1-800-854-0187, 24/7
FAX: +1-512-925-1814

Constituency: Apple Computer Worldwide R&D Community
Response team: Apple Computer Response Squad; Apple CORES
E-mail: Isefton@apple.com
Telephone: +1-408-974-5594
Emergency phone: +1-415-948-5394
FAX: +1-408-974-4754

Constituency: Australia—Internet .au domain
Response team: Security Emergency Response Team (SERT)
E-mail: sert@sert.edu.au

Telephone:	+61-7-365-4417
After hours:	+61-7-365-4417
FAX:	+61-7-365-4477
Constituency:	Department of Defense-interest systems
Response team:	DoD's ASSIST (Automated Systems Security Incident Support Team)
E-mail:	assist@assist.ims.disa.mil
Telephone:	+1-703-756-7974, 9 A.M.–5 P.M., EST
Emergency phone:	+1-800-SKY-PAGE, pin no. 2133937 (pager)
FAX:	+1-703-756-7949
Constituency:	DFN-WiNet, German Internet Sites
Response team:	DFN-CERT (Deutsches Forschungsnetz)
E-mail:	dfncert@cert.dfn.de
Telephone:	+49-40-54715-262
FAX:	+49-40-54715-241
Constituency:	Digital Equipment Corporation and Customers
Response team:	SSRT (Software Security Response Team)
E-mail:	rich.boren@cxo.mts.dec.com
Telephone:	+1-800-354-9000
Emergency phone:	+1-719-592-4689
FAX:	+1-719-592-4121
Constituency:	Energy—U.S. Department of Energy sites and Energy Sciences Network (ESnet)
Response team:	CIAC (Computer Incident Advisory Capability)
E-mail:	ciac@llnl.gov
Telephone:	+1-510-422-8193, 24/7
FAX:	+1-510-423-8002
Constituency:	French Universities, Minister of Research and Education, France, CNRS, CEA, INRIA, CNES, INRA, IFREMER, EDF
Response team:	GIP RENATER
E-mail:	rensvp@urec.fr
Telephone:	+33-44-27-26-12
FAX:	+33-44-27-26-13
Constituency:	Germany—southern area
Response team:	Micro-BIT Virus Center

E-mail:	ry15@rz.uni-karlsruhe.de
Telephone:	+49-721-37-64-22
Emergency phone:	+49-171-52-51-685
FAX:	+49-721-32-55-0
Constituency:	MILNET
Response team:	DDN (Defense Data Network)
E-mail:	scc@nic.ddn.mil
Telephone:	+1-800-365-3642, 8 A.M.–5 P.M., EST
Emergency phone:	+1-703-692-2714, 24/7
FAX:	+1-703-692-5071
Constituency:	NASA—NASA-wide
Response team:	NASA Automated Systems Incident Response Capability
E-mail:	tencati@nssdca.gsfc.nasa.gov
Telephone:	+1-301-794-5201, 8 A.M.–5 P.M., EST
After hours:	+1-759-7243, pin no. 5460866 (pager)
FAX:	+1-301-513-1608
Constituency:	NAVY—U.S. Department of the Navy
Response team:	NAVCIRT (Naval Computer Incident Response Team)
E-mail:	navcirt@nosc.mil
Telephone:	+1-202-282-2601, 7 A.M.–5 P.M., EST
Emergency phone:	+1-800-759-8255, pin no. + 1-2021306 (pager)
FAX:	+1-202-282-0411
Constituency:	NORDUNET—Denmark
Response team:	Nordunet
E-mail:	ber@sunet.se
Telephone:	+46-8-790-6513
FAX:	+46-8-24-11-79
Constituency:	Penn State—The Pennsylvania State University
Response team:	Penn State
E-mail:	krk5@psuvm.psu.edu
Telephone:	+1-814-863-9533, 8 A.M.–5 P.M., EST
After hours:	+1-814-863-4357 or + 1-814-237-8081
FAX:	+1-814-865-3082
Constituency:	Purdue University
Response team:	PCERT
E-mail:	pcert@cs.purdue.edu
Telephone:	+1-317-494-3561
After hours:	+1-317-474-7094

FAX:	+1-317-494-6440
Constituency:	RARE membership
Response team:	Woolwich Centre for Computer Crime Research/CERT Task Force
E-mail:	woolwich@ex.ac.uk
Telephone:	+49-0392-263247
After hours:	+49-0392-436351
FAX:	+49-0392-263247
Constituency:	SBA/Small Business Community Nationwide
Response team:	SBA CERT
E-mail:	bolden@first.org
Telephone:	+1-202-205-6708
After hours:	+1-202-205-6708
FAX:	+1-202-205-7064
Constituency:	SPAN—France
Response team:	SPAN France
E-mail:	harvey%meudon.dnet@east.gsfc.nasa.gov
Telephone:	+33-1-4-507-2805
FAX:	+33-1-4-507-2806
Constituency:	Sprint—U.S. Sprint
Response team:	Sprint DNSU
E-mail:	/PN = DATANETWORK.SECURITY/O = US.SPRINT/ADMD = TELEMAIL /C = US/@sprint.com or /PN = CORPINFO.SECURITY/O = US.SPRINT/ADMD = TELEMAIL /C = US/@sprint.com
Telephone:	+1-703-689-7317, 8 A.M.–5 P.M., EST
After hours:	+1-800-SKY-PAGE, pin no. 44260 (pager)
FAX:	+1-703-689-7380
Constituency:	Customers of Sun Microsystems
Response team:	Sun Microsystem's Customer Warning System (CWS)
E-mail:	security-alert@sun.com
Telephone:	+1-415-688-9080
Emergency phone:	+1-415-688-9081
FAX:	+1-415-688-9101
Constituency:	SURFnet connect sites—Netherlands
Response team:	SURFnet Computer Emergency Response Team
E-mail:	cert-nl@surfnet.nl

Telephone:	+31-30-310290
FAX:	+31-30-340903
Constituency:	United Kingdom—all government departments and agencies
Response team:	CCTA IT Security & Infrastructure Group
E-mail:	sp305@ccta.uk
Telephone:	+44-71-217-3053
Emergency phone:	+44-71-217-3023
FAX:	+44-71-217-3449
Constituency:	Unisys Internal and External Users
Response team:	UCERT
E-mail:	garb@dockmaster.ncsc.mil
Telephone:	+1-215-986-4038
Emergency phone:	+1-215-757-1862
FAX:	+1-212-986-4409

8 Fred Cohen, *The Sciences,* 1991: Sept./Oct.

Chapter

3

Security Standards

"The great thing about standards is that there are so many of them!" ANONYMOUS

"The decisions we make today about communication security will determine the kind of society we have tomorrow." WHITFIELD DIFFIE

[Encryption] Key management developer

Objectives

At the end of this chapter you will be able to:

- Set your own security standards posture in view of government policy.
- Understand the Federal Information Processing Standard (FIPS) philosophy and know where to find FIPS documents.
- Assess the effect of the Computer Act of 1987.
- Discuss nonmilitary computer security standards.
- Understand who the National Security Agency is and their role in setting commercial security standards.
- Weigh the advantages and disadvantages of standard versus proprietary encryption schemes.
- Explain LAN security as standardized in IEEE Standard 802.10B.
- Discuss the International Standards Organization Open Systems Interconnection Reference Model (OSIRM) 7498-2 Security Architecture.

- Compare the Privacy Enhanced Mail, Riordan Privacy Enhanced Mail, Pretty Good Privacy, and Preliminary Message Security Protocols.
- Understand the policy complications of encryption usage abroad.
- See how standards can reduce security.

Security Standards

It seems that today everybody wants to get in on the standards act. What we are seeing in the security field are a few new players plus old players who seem bent on redefining their roles.

This chapter discusses each of the player types and the extent to which they are influencing the development and direction of security technology. It also discusses the major standards and where they fit in the evolution of security standards.

This chapter has not only a technical purpose, to illuminate and give each person perspective, but a social one as well. In a free society, we prize our right to express ourselves as we please. To an equal extent, we cherish our right to privacy when we wish. The success or failure of our society will be based on these two ideals.

Some in our midst choose to test the bounds of expression at both ends of the spectrum. A minority of hackers defend what they do as perpetuating our long tradition of freedom of speech. For instance, they defend posting viruses and virus development tools on bulletin boards. Some insist that accessing information that is not theirs is legal, unless they cause damage or use the information for gain.

At the other end of the spectrum is the need and perhaps even the obsession for secrecy possessed by many organizations in an effort to retain or at least not surrender competitive advantage. They have a clear right to protect what is theirs.

The right of legitimate enterprises to do so is unquestioned. Questions arise when illegitimate endeavors seek to use the security tools of above-board organizations. Here the government steps in, saying in effect, "If you expect us to use our police power to protect the body politic, then you must allow us to conduct surveillance on individuals and organizations that may be working against the public interest."

In an effort to resolve this conflict between the right to privacy and the obligation of government to protect us, we have created a system whereby the government must obtain a court order in order to wiretap a suspected illegal activity. This principle is long established, and court oversight of such taps has worked reasonably well.

Time and technology, however, stand still for no one. Today, criminals have access to the same encryption tools as legitimate businesses. How does the map change when these people start using unbreakable encryption devices? Does the government throw up its hands and quit? Is a change in public policy warranted?

Bearing on this chapter, if we permit standards to be set by the government, what assurance do we have that the government does not possess a way to decrypt anything we encrypt, without court permission? Why should anyone adopt a security algorithm that may contain a built-in back door?

In fact, the federal government is making an effort to assure itself of a back door by proposing that users of new encryption technology post copies of their keys with the government. Divided into two parts and held in two separate places, the holding agencies could give them to law enforcement authorities only under court order. Is this sufficient? Does it represent a logical extension to a wiretap policy that has worked well?

Let us say that we do not care for this idea. If we say that we will use a private, proprietary algorithm, how secure is it, really? Could we not ask the same question of a private vendor? And how do we gauge the resistance of any method to analytical attack? In the end, we have to rely on either government expertise or that of private industry to tell us.

The question is not at all philosophical. Many organizations such as banks have a legal liability to maintain the privacy of their customers. They discharge that fiduciary responsibility in part by using encryption techniques sanctioned by the government, particularly one called DES, the Data Encryption Standard.

Users do not have the time or expertise to research all known encryption methods and pick the most secure one. DES is federally accepted and a documented standard enshrined in Federal Information Processing Standard (FIPS) Publication No. 46-1 and ANSI X3.92-1981/R1987. If users deploy DES properly, the auditors will likely be satisfied.

So we come back to standards again. At least a standard is a known quantity. At least we will not adopt a standard without understanding its strengths and weaknesses. We understand the balance between public and private standards and that they have a legal dimension that we cannot ignore.

Who Is Making the Standards?

As Table 3.1 shows, the following groups make standards:

TABLE 3.1 Standards Makers

Standards maker	Encryption	Authentication	TEMPEST emissions	Policy and procedure	Trusted systems
Clinton admin.	Clipper, Capstone	Same		NSC review	
NIST	FIPS 46-1	FIPS 113		FIPS 48, 83, 112, 74, 31, 41, 73, 102	
FTSC	DES			With NIST	
ANSI	DES X3.92	X9.9			
NSA	Capstone, Clipper, Skipjack, DES	Digital Signature Standard	Yes	NCSC, TG Series, Red Book, Orange Book,* and others	Same
IEEE	IEEE 802.10B	IEEE 802.10B			
ISO	ISO 7498-2	ISO 7498-2			ISO 7498-2
IAB	Uses DES	Uses RSA†			
Vendors	IBM	RSA			
CSSPAB				Yes	

*The National Center for Computer Security, a government entity, publishes this series of unclassified security references.
†RSA denotes the Rivest–Shamir–Adelman public-key technique.

- The Clinton Administration
- The National Institute of Standards and Technology (NIST)
- The Federal Telecommunications Standards Committee (FTSC)
- The American National Standards Institute (ANSI)
- The National Security Agency (NSA)
- The Institute of Electrical and Electronic Engineers (IEEE) 802.10B LAN Security Working Group [Standard for interoperable LAN security (SILS)]
- The International Standards Organization (ISO)
- The Internet Architecture Board (IAB)
- Indirectly, the vendor community
- The Computer Systems Security and Privacy Advisory Board (CSSPAB)

What Do These Security Standards Do?

The standards fall into these functional categories:

- *Encryption* involves scrambling information into pseudorandom form so that only a recipient with the correct key can unscramble it.
- *Authentication* is accomplished by adding a "signature" to the end of a document or file to:

  Prove that the sender is really the originator

  Prove that the document or file has not been tampered with

  Make sure the receiver is the intended receiver

  Make sure the receiver really did get the message and so cannot repudiate it later

- *Emissions control* is accomplished by shielding or isolating electromagnetic emissions so that they cannot be received by unintended recipients. This process is commonly called TEMPEST (not an acronym) and is now synonymous with *compromising emanations,* or emanations that compromise information security.
- *Policies and procedures* should be spelled out in standards that enhance security, such as password lengths and password-change rules.

- *Design of trusted systems* involves the creation of computer systems with security controls and audit paths, yielding a formally specified level of security.

From a communications security standpoint, our interest is focused on the first three areas, with an interest in trusted systems to the extent that communications is involved. Some of the tools used in trusted systems such as audit trails are valuable in providing network security, too.

The Clinton Public Encryption Management Directive

In April 1993 the Clinton Administration, citing the need for improved security and privacy in communications systems, called for implementation of a new encryption technology called Clipper (the name of an encryption chip set).

The President asked the National Institute of Standards and Technology to create a federal standard based on Clipper within six months. The standard would make Clipper a mandatory procurement provision for unclassified but sensitive federal applications. Keys to the Clipper system would be kept by the government in escrow. The President cited the need to balance individual privacy rights with law enforcement and security concerns in justifying the escrowed keys.

At the same time, a classified presidential directive requested the federal government to review its policy on encryption, privacy protection, and law enforcement. Managed and directed by the National Security Council, the review could result in further changes or even restrictions to communications technology. The directive did not publicly state its exact scope or the review procedure(s) to be followed. An interim report was due in June 1993 and a final report by August or September of that year. It is not known whether these dates were met.

Implications

The twin directives sent the security community into hysteria, both from a technical and a policy viewpoint. Technically, the objections raised were the following:

- The Clipper chip is proprietary, made by one vendor. There has been no public validation of its integrity or quality. Said Gilles Brassard, a mathematics professor at the University of Montreal,

"It is a giant leap backwards to rely on a cryptosystem whose algorithms are kept secret." Peer review is the only way to test for validity.

- Neither has the Skipjack encryption algorithm buried in the Clipper chip received public scrutiny.
- Keys would be stored by the government in escrow and would be released only under court order.
- With this built-in back door, users would likely avoid Clipper, preferring the older but trusted DES. This would be especially true for criminals.
- Back doors to Skipjack might exist that do not require the use of the escrowed keys, although a panel of experts has said there is no such back door.
- Once revealed, the keys can be used to eavesdrop indefinitely, beyond any time limit set by court order.
- Clipper does not comply with European bank security standards.

There were many policy objections, too, as follows:

- There has been no open debate about any aspect of such a system.
- With the government's purchasing power and stature as de facto and de jure standards maker, Clipper could drive other security vendors out of business.
- The policy review might result in restrictions on other forms of encryption in the United States. Currently, there are no such constraints.
- Some people claim that the National Security Agency, who designed the Skipjack algorithm, has no real interest in commercial security. It makes their eavesdropping job that much harder.
- Charles Cresson Wood, a noted security consultant, said in a letter to *Network World*, "We need a national debate about the implications of such a technology, not back room deals among government spooks."

Industry response has been vigorous. The Electronic Frontier Foundation (EFF), a coalition of about 40 vendor and civil liberty organizations, has made its concerns known to the Clinton Administration. EFF was founded by Mitchell Kapor, of Lotus fame. He said, "A system based on classified, secret technology will not and should not gain the confidence of the American public."

The Clinton Administration said in June 1993 that it adopted a "go slow" approach in view of the resistance encountered from corporate users. In July 1993, an industry group headed by Novell met in San Francisco and announced their intention to offer a set of standards to protect information stored in computers or transmitted over networks.

The group's announcement amounted to an indirect challenge to the White House. However, since the government cannot at this time impose its own standard or block development of an industry standard, the administration again took the soft approach. Said one official, "We'll look at this and evaluate it on its technical merits."

The industry proposal is initially based on NetWare and would use Rivest–Shamir–Adelman (RSA) public-key techniques plus DES encryption. The group proposed that instead of encrypting data on an ad hoc basis, an industrywide standard should be used in all computer networks.

RSA and DES are strictly regulated export items, regulated even more closely than parts for nuclear weapons. For the proposal to succeed, the government would have to reclassify this technology as nonstrategic. The government's concern is that the Saddam Husseins of the world will use our encryption technology to hide their evil machinations. What really happens is that they simply buy whatever they need elsewhere.

The government's proposal is a good start, but is conceptually flawed. First, the Clipper chip is aimed at voice, not data communication. Second, Congress could pass a law making nongovernment encryption schemes illegal (although such a law would surely be challenged in court). Third, the government's stature is such that anything it approves officially becomes a standard if it is used enough.

One *good* thing about Clipper is that it would make ordinary voice conversations and dial-up data exchanges much more secure. Last year the FBI caught some 400 illegal phonetappers. The next logical question is, how many did they miss?

The National Institute of Standards and Technology

Formerly the National Bureau of Standards, NIST was created in 1987 and is part of the Department of Commerce. NIST issues the Federal Information Processing Standards (FIPS). Of the many FIPS published, 16 have to do with security. Among them are FIPSPUB 46-1, the Data Encryption Standard.

FIPS publications are often cited as references in government and private procurements. The Computer Systems Laboratory of NIST is responsible for "developing technical, management, physical and administrative standards and guidelines for the cost effective security and privacy of sensitive information in Federal computer systems." In reality, NIST-developed standards are widely used in industry as well. Adoption of a FIPS standard makes federal procurement of products and services based on the applicable FIPS mandatory.

The American National Standards Institute

NIST standards and those of the American National Standards Institute (ANSI) occasionally overlap. For instance, FIPSPUB 46-1 is one of many FIPS adopted by ANSI, in this case, renamed ANSI X3.92.

ANSI is the United States' representative to the International Standards Organization (ISO). Standards proposed to ISO must first pass through ANSI. Many groups make such submissions, including NIST, the Institute of Electrical and Electronics Engineers (IEEE), and others.

The National Security Agency

On matters of security, NIST works closely with the National Security Agency (NSA). NSA is responsible for electronic eavesdropping or "technical intelligence gathering" and government and/or military information security. Supersecret, NSA is colloquially dubbed the "Puzzle Palace" and is based at Fort George G. Meade in Maryland.

NSA approved the Data Encryption Standard, which was developed by IBM. NSA developed the Skipjack algorithm to be used in Clipper and an associated system called Capstone.

NSA operates the National Computer Security Center (NCSC) to help private industry develop "trusted" computer products through publication of guidelines (see Table 3.2) and a Trusted Product Evaluation Program in which NCSC evaluates private systems. This is a rigorous and consequently expensive process. A vendor team led by Novell to certify a secure version of NetWare at the C2 level (a fairly low security level) is expected to take two years and cost $17 million.

TABLE 3.2 NCSC "Orange Book" Evaluation Divisions and Classes

Division	Protection	Class	Protection	Comments
D	Minimal	D	Minimal	Evaluated, but does not meet the requirements of any higher class.
C	Discretionary	C1	Discretionary security	Confidence in hardware and software controls. Isolates users and data. Ensures authenticity of users and data. Many commercial OSs meet C1.
		C2	Controlled access	Encapsulated resources; login and explicit auditing. Add-on security software often meets C2. Per a Department of Defense directive in 1992, most computers handling government unclassified or sensitive unclassified data must meet C2. Novell is seeking C2 in NetWare 4.X; C2 security is also planned for Windows NT.
B	Mandatory	B1	Labeled security	Explicit protection model required; execution domains; file labels; system security officer and documentation required.

		B2	Structured	Formal security model required; kernelized; covert channel ID required; tracing; mandatory controls including communications lines. Examples: Multics-AIM; Verdix' Multi-Level Secure/LAN; Windows NT (planned).
		B3	Security domains	Central encapsulation; reference monitor; tamperproof; recovery procedures; protected against authentication attacks.
A	Verified	A1	Verified design	Extensive security consideration during all development phases. Math tools; formal models with explicit security theorems; formal top-level specification; trusted software distribution. Example: Honeywell SCOMP operating system.
		Beyond class A1		Developmental; source verification.

The IEEE 802.10B LAN Security Working Group

The IEEE is the Institute of Electrical and Electronic Engineers. IEEE 802 standards turned Ethernet into a nonproprietary standard called IEEE 802.3 and IBM's token passing–token ring scheme into IEEE Standard 802.5.

It therefore makes sense that IEEE would have a group addressing LAN security. As Fig. 3.1 shows, the standard for interoperable LAN/MAN security (SILS), IEEE Standard 802.10-1992 defines an ISO layer 2 security protocol that can be used to protect IEEE 802 LANs and metropolitan area networks, (MANs). The standard provides confidentiality and assurance of data integrity. When used with key management or systems management, authentication of the origin of data and controlled access are supported.

The Secure Data Exchange (SDE) protocol is the main subject of IEEE Standard 802.10. SDE specifies services that permit secure data exchange at the data link layer of the ISO model. SDE is sandwiched in between the media access control (802.3-6) and logical link control (802.2) sublayers.

Key management allows electronic management of the cryptographic keys that encrypt data at layer 2. The vehicle is the Key Management Protocol that operates at the application layer. Also at the application layer is the System/Security Management Protocol. This protocol consists of services used to manage the other security protocols in such a way that they cannot be breached.

SILS relies heavily on the ISO/OSI Reference Model, ISO-7498, and the Security Architecture Addendum, ISO-7498-2.

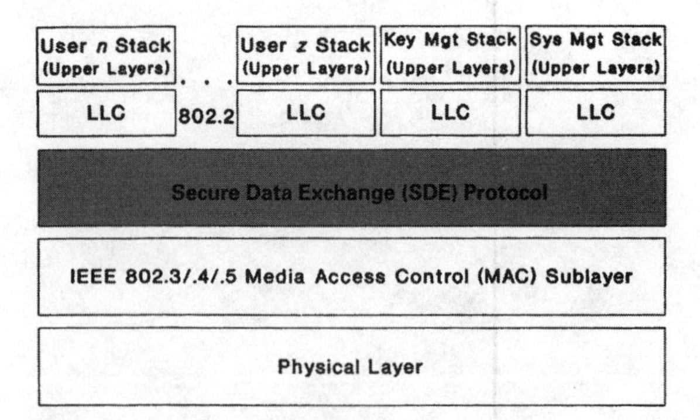

Figure 3.1 SILS in the ISO Reference Model. (LLC = logical link control.)

The International Standards Organization

ISO is *the* big gun in making standards. Founded in 1947 and located in Geneva, Switzerland, ISO builds standards for many activities including meteorology and photographic film speeds (see your 35 mm cassette where it says, e.g., "ISO 200" speed).

Standards are proposed to ISO via member country-based bodies. In the United States, ANSI is the accredited representative to ISO. The British Standards Institute in the United Kingdom is a representative, as the Canadian Standards Institute represents the Dominion of Canada.

For security, the applicable document is ISO 7498-2, published in 1989. It is an addendum to the well known seven-layer ISO Reference Model described in ISO-7498. ISO 7498-2 defines 14 security services, some or all of which may be available at a given layer. As Table 3.3 shows, at the application layer, all are available, but there are no security services available at the session layer. These security services are defined in Table 3.4.

Since we work heavily at layer 2, both connection (oriented) confidentiality and connectionless confidentiality are available. At layer 1, only the former and traffic flow confidentiality are offered.

The Addendum recognizes that not only the network but its management may be attacked. It addresses specifically the security of the interface between a host (or server) and the external universe.

All, some, or none of the services listed in Table 3.3 may actually be implemented at a given layer. Such services may appear at all, some, or none of the ISO layers. The chart means that ISO 7498-2 recommends the services shown be made *available* as a provider option.

Although no services are specified at the presentation layer, ISO 7498-2 notes that the presentation layer contains a number of security *facilities* (italics theirs) that support the provision of security services by the application layer.

Access control is regarded as the key. It consists of two parts: claiming an identity and then proving it via authentication. An entity seeking admittance must state its capabilities or its credentials from which the admitting system will allow or disallow access. If allowed, the admitting system retains a responsibility to enforce any access limitations attached to the entity.

TABLE 3.3 OSI Security Services by ISO Model Layer*

Security service	A	P	S	T	N	D	P
Peer entity authentication	Y	N	N	Y	Y	N	N
Data origin authentication	Y	N	N	Y	Y	N	N
Access control service	Y	N	N	Y	Y	N	N
Connection confidentiality	Y	N	N	Y	Y	Y	Y
Connectionless confidentiality	Y	N	N	Y	Y	Y	N
Selective field confidentiality	Y	N	N	N	N	N	N
Traffic flow confidentiality	Y	N	N	N	Y	N	Y
Connection integrity with recovery	Y	N	N	Y	N	N	N
Connection integrity without recovery	Y	N	N	Y	Y	N	N
Selective field connection integrity	Y	N	N	N	N	N	N
Connectionless integrity	Y	N	N	Y	Y	N	N
Selective field connectionless integrity	Y	N	N	N	N	N	N
Nonrepudiation, origin	Y	N	N	N	N	N	N
Nonrepudiation, delivery	Y	N	N	N	N	N	N

*A = application, P = presentation, S = session, T = transport, N = network, D = data link, P = physical.

TABLE 3.4 OSI Definitions of Security Services

Security service by ISO model layer	ISO 7498-2 definition (expanded)
Peer entity authentication	The corroboration that a peer entity in an association is the one claimed. This service, when provided by a layer, corroborates to the next higher layer that the higher layer elsewhere is the one it claims to be.
Data origin authentication	The corroboration that the source of data received is as claimed. This service, when provided by a layer, corroborates to the next higher layer that the source of the data is the claimed (peer) higher-layer entity.
Access control service	Access control: The prevention of unauthorized use of a resource, including the prevention of use of a resource in an unauthorized manner. This service protects against unauthorized use of resources accessible by OSI.
Connection (oriented) confidentiality	Confidentiality: The property that information is not made available or disclosed to unauthorized individuals, entities, or processes. This service provides for the confidentiality of all data at a given layer for its connection to a peer layer elsewhere.
Connectionless confidentiality	Same as above, except for data at a given layer in a single connectionless service data unit (SDU).
Selective field confidentiality	This service provides for the confidentiality of selected fields within the user data on a connection or in a single connectionless SDU at a given layer.
Traffic flow confidentiality	A confidentiality service to protect against traffic analysis (this means traffic presence, absence, amount, direction, and frequency).
Connection integrity with recovery	This service provides for the integrity of all user data on a connection at a given layer and detects any modification, insertion, deletion, or replay of any data within an entire SDU sequence (with recovery attempted).

TABLE 3.4 OSI Definitions of Security Services (*Continued*)

Security service by ISO model layer	ISO 7498-2 definition (expanded)
Connection integrity without recovery	Same as above, but with no recovery attempted.
Selective field connection integrity	This service provides for the integrity of selected fields within the user data of a given layer's SDU transferred over a connection and takes the form of determination of whether the selected fields have been modified, inserted, deleted, or replayed.
Connectionless integrity	This service, when provided by a given layer, provides integrity assurance to the requesting entity in the layer above it. This service provides for the integrity of a single connectionless SDU and may take the form of determination of whether a received SDU has been modified. Additionally, a limited form of detection of replay may be provided.
Selective field connectionless integrity	This service provides for the integrity of selected fields within a single connectionless SDU and takes the form of determination of whether the selected fields have been modified.
Nonrepudiation with proof of origin	The recipient of data is provided with proof of the origin of data. This protects against any attempt by the sender to falsely deny sending the data or its contents.
Nonrepudiation with proof of delivery	The sender of data is provided with proof of delivery of data. This protects against any subsequent attempt by the recipient to falsely deny receiving the data or its contents.

The Federal Telecommunications Standards Committee

FTSC is an interagency advisory body. They work to

- Help telecommunications networks interoperate
- Assist NIST in setting data communications standards
- Ensure that the federal government participates in setting standards

FTSC works closely with ISO, ANSI, the International Telecommunications Union–Telecommunication Standardization Sector (ITU-TSS), and the Electronic Industries Association (EIA) to adapt or apply existing standards to federal use. FTSC generally avoids creating their own standards, but they did make an exception and develop criteria for implementing DES when DES became a federal standard in 1979.

The Internet Architecture Board

As manager of the Internet, the world's largest computer network, the IAB has oversight responsibility for what happens within the network. The network, originally created by the Department of Defense's Advanced Research Projects Agency, is used extensively by military, industrial, and commercial organizations plus private users. At the detail level, the IAB has the Internet Engineering Task Force (IETF) to advise it on short-term technical matters and establish protocols to be used within the Internet.

Privacy Enhanced Mail (PEM) is the application standard for encryption in the Internet. PEM is intended for use with the Simple Mail Transfer Protocol (SMTP). PEM uses RSA public-key and DES algorithms, although PEM is designed to be able to be used with a variety of encryption algorithms. To implement PEM, the Internet Society plans to administer a hierarchical system to certify public keys and so assure that the keys used are authentic.

RIPEM (standing for RIordan PEM) is a subset of PEM. RIPEM does not use certificates. Since the public keys are unsigned, the user is left to make his or her own assessment of the validity of a key. By this time, there may be an enhancement to permit the signing and hence the validation of key certificates. RIPEM uses RSA public keys and so is subject to RSA patent license provisions.

The military equivalent of PEM is the Message Security Protocol (MSP). MSP was developed by NSA in the late 1980s through the Secure Data Network System program. NIST and the vendor community also participated. MSP is an X.400 electronic-mail-compatible application-layer protocol designed to protect e-mail.

MSP is a generic protocol that supports a variety of security functions, including signing (adding a validating signature, as on a letter), hashing (making sure a message was not tampered with), and encryption. NIST has proposed that MSP be adopted as an ISO standard. NSA is also proposing that the IETF use ISO MSP plus the Digital Signature Standard, which we will discuss in Chapter 4.

A variation of MSP, called Preliminary MSP (PMSP), is designed for nonmilitary but sensitive messages. It can be used with the X.400 system or TCP/IP's SMTP. PMSP was finalized in December 1992 and titled Mosaic (not to be confused with the Internet browser of the same name) in April 1993. It will work with the Capstone public-key chipset.

Like PEM, both MSP and PMSP can support a variety of encryption algorithms. Both PMSP and MSP can support seamless secure message exchanges between SMTP and X.400 systems, but PEM cannot.

Pretty Good Privacy

Pretty Good Privacy (PGP) was developed by Philip Zimmermann in 1992 outside the United States to avoid the legal strictures imposed by the International Traffic in Arms Regulations rules and the RSA patents that exist only in the United States. At this writing, there is a pending federal indictment against Zimmerman for having exported PGP outside the United States. Apparently PGP was posted on the Internet domestically and then copied overseas without his knowledge. For practical purposes, current and potential users of PGP in the United States should know that their right to use PGP is under a legal cloud. This has not stopped many users of PGP. A rather lively discussion of PGP occurs regularly on the Internet's USENET under comp.security.misc.

PGP is a combination of an encryption algorithm called IDEA, a public-keying structure, and a hashing algorithm called MD5 worked out by RSA co-developer Ronald Rivest in 1992. PGP does not use DES, but another structurally similar encryption algorithm called IDEA. Like DES, IDEA is a block cipher, meaning that both algorithms operate on a block of bits. DES, for example, turns 64-bit blocks of plain text into ciphered 64-bit blocks. IDEA, through a patented algorithm, can be used without charge for noncommercial use. IDEA is a much newer algorithm than DES (1992 versus 1978) and uses a 128-bit key as opposed to DES's 64 bits including eight parity bits. However, a new algorithm in the cryptography community is more suspect than an older one that has been scrutinized exhaustively for weaknesses. IDEA is still under the microscope even as it is being used.

Aside from secrecy, PGP includes message hashing to assure the integrity of a message. This is accomplished using an algorithm called MD5.

Users of PGP obtain signed packages of keys from other users. When added to other packages received from other users they form what is known as a *keyring*. As with RIPEM, the user determines the extent of the trust placed in the signer, and so decides indirectly the degree of faith in the key package. Forgers can contaminate the trust tree (higher keying levels) at the top of the structure or anywhere. As a result, PGP users must go up the line and assure themselves of sufficient trust with users upstream with whom they wish to communicate. The simple existence of a tall tree is meaningless from a security standpoint.

The Vendor Community

A small number of vendors are involved in creating security standards. Often, they become involved as contractors to the government. Other times, independent researchers develop an algorithm and then establish an organization to capitalize on it.

IBM developed the Data Encryption Standard in 1977, under contract to NSA. IBM grants a nonexclusive royalty-free license to anyone who wishes to use DES. DES is alive and well today. Numerous attempts to crack it have failed. With 1994 technology, the time required to break DES by a brute-force attack is estimated at ten years.

Even so, DES's life is easily extensible simply by making the key longer: today it is 56 bits plus 8 parity bits; doubling it to 112 bits plus 16 parity bits would give it many years' more life. NIST has reconfirmed DES as a standard until 1998.

RSA Data Security, Inc. (Redwood City, Calif.) is a company formed by the discoverers of the public-key technique. The principle behind the RSA technique is that if one takes two long prime numbers and multiplies them together, finding the two numbers from the result is very, very difficult.

Ronald Rivest, A. Shamir, and L. Adelman, from the Massachusetts Institute of Technology and Stanford University, wrote a paper about this technique in 1978 and obtained patents for the process. Public-Key Partners, Inc. owns the rights to the RSA technology. RSA exploits the patents and does not grant royalty-free licenses as IBM does for DES. (However, there is a free but still licensed version of the code on the Internet called RSAREF.) Their fee-for-license policy has made RSA technology less desirable for adoption into standards, especially by the U.S. government.

The Novell NetWare security patches use RSA authentication. RSA is also used by the international SWIFT banking network. RSA encryption algorithms are referenced in international standards (CCITT X.509, an international security standard; ISO 9796;

the Internet's PEM) and widely licensed. Needless to say, RSA opposes the Clipper initiative that could put RSA out of business.

RSA maintains an Internet FTP site titled rsa.com (alias: chiralty.rsa.com). In it are articles on the Clipper chip; frequently asked questions (FAQs) and question and answer dialogues; publications; and a selection of Internet requests for comments (RFCs).

Mykotronx, Inc. (Torrance, Calif.) is the only NSA-approved chipmaker for the Clipper and Capstone chipsets. Both chips were developed in concert with NSA for use in the military Defense Message System.

Motorola, Inc. (Scottsdale, Ariz.) has a long history of building encryption devices, the best-known of which is the STU-3 Secure Telephone.

Motorola's principal adversary for security products is *AT&T,* which has backed the development of Clipper and Capstone. AT&T offers its own secure telephone accessory using Clipper chips; it is called the AT&T Telephone Security Device.

The Computer Security Act of 1987

The Act took nonmilitary security standards development away from NSA and gave it to other civilian agencies, such as NIST. The Act's purpose was to limit the role of NSA in developing security standards by subjecting nonmilitary computer standards to public review. The Act also established penalties for most hacking activities.

In reality, NIST fronts for NSA. While NIST's signature may be on the documents, NIST relies heavily on NSA for expertise. Indirectly, NSA exercises yea-or-nea control over many nonmilitary security decisions.

National Computer Systems Security and Privacy Advisory Board

Created by the same Computer Security Act, the NCSSPAB consists of 12 industry, consulting, and government security experts. They advise Congress on security and privacy issues and identify issues of national policy importance. They are an independent (but toothless) watchdog over federal security activities.

In 1994 NIST planned to replace its current guidelines for evaluating the security aspect of information systems. The old guidelines (NCSC's "Red Book" and "Orange Book") focused too much on computer operating systems and not enough on networks and databases. In September 1993, NIST outlined these problems to CSSPAB and presented their plans to add balance to the guidelines.

However, this was an indirect courtesy to Congress and not obligatory. Nor is NIST obliged in any way to take any action on feedback given by CSSPAB.

NIST is developing these new guidelines, called the Common Criteria, with input from Canada, Japan, and Europe. The guidelines will be proposed through ANSI as a standard to the ISO. The result will be an international market in "trusted systems," including distributed systems, networks, encryption, and PC security.

Another change will be in the way certification is done. Currently NSA evaluates all systems for certification. This has proved expensive and time consuming (e.g., a five-year cycle). In Europe, independent testing laboratories do much of this work. That model is being proposed here, except for the high-level systems that NSA will continue to certify.

One criticism aimed at NIST by CSSPAB is that NIST has not involved the user community in the early stages of this new policy development. It is an exact repeat of the Clipper controversy in which the user community was handed a *fait accomplis.*

Transborder Encryption Issues

International

In many countries, it is illegal to use a device to encrypt data crossing a national border. Some do not even allow encryption within their borders.

In other nations, the use of encryption is allowed, as in Japan, but the government must hold a copy of the keys. There is no assurance that the keys will not be used to eavesdrop, and as we have noted, the keys *are* used to decrypt traffic at the government's sole discretion.

To turn the circumstance around, how would foreign governments feel about the U.S. government holding keys to Clipper chips used in their respective countries? Thus the Clipper controversy adds an international dimension. Furthermore, no one overseas, not even our allies, was asked about Clipper. In the words of Ross Anderson, editor of the British publication *Security Reviews,* "I believe the Clipper will be useless as a means for protecting U.S. business communications with foreign subsidiaries and partners. There was no coordination with the allies on the Clipper announcement, so there is little good will."

There could be a negative effect on exports of American telecommunications equipment with Clipper chips inside. First, the State

Department may not allow it. The State Department controls exports of current encryption technology through the issuance or denial of export licenses. Why should Clipper be different?

Both the Industry Coalition on Technology Transfer, a coalition of 10 industry trade associations, and the U.S. Council for International Business, a group of 290 banks, manufacturers, and telecommunications vendors, have criticized the Clipper chip specifically and the government's export policy in general.

Domestic

Under pressure from the Department of Defense, the U.S. State Department's Office of Defense Trade Controls licenses export of encryption hardware and software as though they were munitions under the International Traffic in Arms Regulations (ITAR). Accordingly, encryption tools are listed on the U.S. Munitions List as a restricted export. NSA is said to routinely recommend denial of licenses, even to friendly countries.

Bulk encryption tools attract most of the attention; encryption utilities in programs to protect passwords are not so much the issue. However, some large companies such as IBM have a rather conservative policy requiring employees to leave any software using DES encryption out of any machines traveling with them abroad. This rule is also the reason you may see "Licensed for USA Use Only" on so many computer products. Software and hardware deemed less strategic fall under less restrictive Department of Commerce rules.

This policy has been obsolete for years. Encryption software is passed easily through the Internet (for example, PGP) all over the world, or simply via a dial modem call and a file transfer protocol. The government's antiquated policy is a case of locking the barn door after the horse has fled.

Domestic vendors claim the policy of equating encryption equipment to munitions simply opens the door for foreign vendors. The result is that U.S. vendors have suffered by having to support multiple versions of hardware and software. Certainly, the loss of international business has pushed domestic prices higher. Corporate multinational networks have been crippled as well as bids for foreign network contracts by American firms.

Licensing can take months. The process is easier if the customer is a financial institution, but otherwise the company must be at least 50 percent American-owned. By one estimate, 40 percent of the international market is thereby eliminated for American vendors.

These concerns have been aired to CSSPAB and are part of the review being performed by the National Security Council. However, in view of the recent Clipper communiqué, few expect the export policy to change.

Do Standards Reduce Security?

If the security standard is designed to be compromised, then the standard reduces security. There is no way to know if such a back door exists without peer review. Therefore, if the security standard has not been subjected to peer review, the "standard" remains suspect.

If an organization (NSA) pursues its own agenda that conflicts with the needs of industry and commerce, then the needs of industry and commerce will inevitably be met in another way. It simply makes no sense to adopt a standard that may be flawed to begin with. The very presence of a key escrow system compromises such a system's security. It is better to use a known-secure method such as DES.

DES has been scrutinized via peer review more than any other encryption algorithm. No real weaknesses have been found. As was mentioned earlier, peer review by disinterested parties is the core of a secure algorithm.

As any standard ages, it becomes increasingly vulnerable. Analytical power increases. With time, subtle weaknesses may be found. If cracked, vast numbers of systems are compromised, giving cryptanalysts strong motivation. A standard implies a static state, while technology advances dynamically. A standard is by nature a sitting duck.

Nonstandard algorithms are not subject to peer review as are proposed or official standards. Proprietary systems may or may not contain weaknesses. However, they are not attacked as often since the broad potential benefit is not as great. Their nonpublic nature makes information about them hard to get, and they can be strengthened at will. Is it therefore an advantage to purchase a standards-based system whose weaknesses are at least known?

For all these reasons, and although it bears watching, DES remains the encryption scheme of nongovernment users' choice for the future.

Review

- The standards we set governing encryption do and will influence our society and our civil liberties.
- Government is making an effort to standardize encryption via the Clipper and Capstone technologies.

- Clipper and Capstone are secret encryption algorithms created by NSA.
- Both give the government a back door around encryption.
- Keys would be held by two or more federal agencies and would be accessible to law enforcement only by court order.
- There are many objections to this plan, based on both technology and policy.
- The right to privacy and the obligation of government to protect us are in conflict.
- There are about ten makers of standards.
- Standards address encryption, authentication, electromagnetic emissions, policies and procedures, and the design of trusted systems.
- The National Security Agency and the National Institute of Standards and Technology are the principal government makers of standards.
- NIST publishes the Federal Information Processing Standards that are mandatory in government purchases and are widely used by private organizations.
- NIST was created by the Computer Security Act of 1987. The Act specified that NSA would advise NIST. Security plans are to be reviewed by NIST, NSA, and the Office of Management and Budget. The Act was an effort to reduce NSA's power outside military circles.
- Some view Clipper and Capstone as an effort by NSA to reassert its authority.
- Policies vary widely regarding the use of encryption in foreign countries.
- Domestic licensing issues have made multinational secure networks harder to build.

Appendix. Contact List

The Clinton Administration
Reachable by mail, MCI Mail via
White House Bulletin Board or at
Internet address: president@
whitehouse.gov

The National Institute of Standards and Technology (NIST)
National Technical Information Service: (703) 487-4650 (Ordering)
Information: (703) 487-4600
5285 Port Royal Road,
Springfield, Va. 22161
Request FIPS Publications or Publications List 58

The Federal Telecommunications Standards Committee (FTSC)
General Services Administration
Specification Distribution Branch
Building 197 (Washington Navy Yard)
Washington, D.C. 20407

The American National Standards Institute (ANSI)
The International Standards Organization (ISO)
(212) 642-4900
11 West 42nd St.
New York, N.Y. 10018
ISO standards can also be ordered from ANSI

The National Security Agency (NSA)
Source for Colored Books
(301) 688-4657
National Computer Security Center
Fort George G. Meade, Md. 20755-6000

The Institute of Electrical and Electronic Engineers
IEEE Service Center
(800) 678-IEEE or (908) 562-3800
445 Hoes Lane
P.O. Box 1331
Piscataway, N.J. 08855-1331

The Internet Architecture Board
Reston, Va.

Chapter

4

Authentication Methods

Objectives

At the end of this chapter you will be able to:

- Define authentication and explain the forms it can take
- Explain what authentication can and cannot do
- Define the principles of *repudiation* and *nonrepudiation*
- Explain what is meant by a *signature*
- Differentiate authentication applications from those requiring encryption
- Show how authentication and encryption can work together
- Understand the utility of authentication in Electronic Data Interchange (EDI), file and program applications, and elsewhere
- Explain what a *key* is, what it does, and why it is needed
- Discuss one-key versus two-key encryption and authentication systems
- Understand what to look for in authentication systems
- Explain what public-key signatures do and describe the elements of the Digital Signature Standard
- Describe how dial-back systems work
- Analyze the weaknesses and strengths of dial-back
- Discuss non-dial-back authentication options

Authentication

What do we mean when we say this word? *Authentication* is the process of *determining* accuracy, the accuracy of:

- The sender's identity, or *data origin authentication*
- The sender's message, or *data integrity*
- The receiver's identity, or *peer entity authentication*

In short, data origin authentication corroborates the source of the data with the claim of the sender. Data integrity ensures that data have not been *altered* or *destroyed* in an unauthorized manner. Peer entity authentication corroborates a peer entity in an association as the one claimed. The corroboration can be either unilateral (the sender or receiver only) or mutual (both sender and receiver).

A Two-Part Process

Authentication is really a two-part process. First the sender, requester, or initiator *identifies* him- or herself. The identification is a statement of who the user is.

You begin the authentication process every time you log in by entering your user ID. You have identified yourself. You must now prove that you are really who you say you are. You do this through authentication.

So authentication is *proof of identification,* or the process by which a claimed identity is verified. Most often, it is your *password* that verifies your claim. Identification is easy; there is nothing secret about it. The hard part is authentication. There are three ways to verify a user's identity (see also Fig. 4.1):

- *By something they know.* Users know passwords; passphrases, personal identification numbers (PIN numbers, as used with an ATM card); data about themselves and their families.
- *By something they have.* Users may possess physical or electronic keys, magnetic-strip cards or badges, or ATM bank cards. You can see then that using an ATM card is a form of double authentication.
- *By something they are.* No two human beings are the same. We use this fact of nature to tell ourselves apart. Examples include fingerprints, retinal patterns, DNA patterns, and hand geometry.

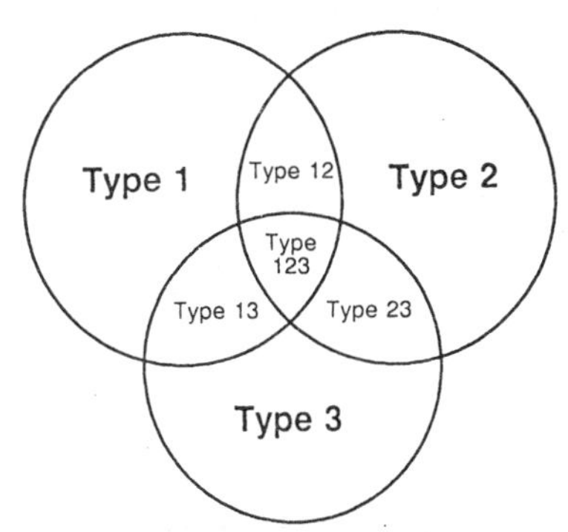

Figure 4.1 Methods of authentication. (*Source: National Security Agency / National Computer Security Center.*)

Authentication mechanisms must uniquely and unforgeably identify an individual. Possession of knowledge or a thing means that it could be given (or taken) by someone else. Since there is no physical attachment of the person to the knowledge or thing, there is an opportunity for loss, theft, or duplication.

There is no such opportunity if we measure something the individual is. Whether a fingerprint, hand size, or retinal pattern, we can uniquely identify that person. This is called *biometric* authentication.

An interesting twist is in the case of identical twins. Here a DNA test would be exactly the same. Other means might not be sufficiently accurate, either. We would have to look to some form of possessive identification in order to tell the twins apart.

Authenticating can use one or more of these techniques. As we add more checks, the statistical certainty of authentication goes up, but so does the cost. We must therefore make judgments as to how much the additional protection is worth relative to the risk.

There is also the matter of *granularity*. Perhaps it is not necessary to identify you individually. If you are an employee, that might be sufficient. When you enter your place of work each day, simply showing a badge may be adequate. Your name is not required.

Perhaps an extreme level of certainty is required. Then all three methods might be used. In real life, though, we rely heavily on knowledge and possession, as of a password and a badge. Except for extreme security requirements, biometric systems are too expensive.

Figure 4.2 An electronic signature.

How Authentication Works in Electronic Systems

Since authentication is a process, it works exactly the same way in electronic form as it does for "liveware" (us). In electronic form, a message is sent with an extra unit of information, usually at the end of the message. It is called a *signature* or a *digital signature,* just as in a letter. (See Fig. 4.2).

Using a letter as an example, the signature definitively identifies the sender in a unique way to the receiver. The receiver knows the message is from the sender, but no more. Nor does the receiver know if the information was tampered with or perhaps even destroyed.

The sender has no way of knowing if the message was received, nor even if the right receiver got it. A letter is an example of data origin authentication using something the sender and receiver know: the signature at the bottom of the letter. Of course, the presumption here is that the receiver had prior knowledge of the relationship of the signature to the sender.

This is a characteristic of all authentication and encryption systems. Something starts the process, saying that a particular signature goes with a particular sender. From then on, we leave it to the machines. If you think about this, you will see it is true. When you log in to a system for the first time and are asked for a password, you establish a relationship. A human being authorizes it; all the machine does is perpetuate it.

Higher forms of signatures exist. We have already discussed one: the cyclical redundancy check (CRC) that is computed based on the data in a file and attached at its end as a way of detecting a virus. The CRC verifies *data integrity,* but nothing else. We do not know where the file came from, or where it might be going. We only know that it has not been tampered with, as by a virus or by a hacker.

Signatures can authenticate:

- The sender to the receiver
- The receiver to the sender

- The integrity of the data
- The transmission and reception of the message (confirmation)

With confirmation, we then say that the receiver cannot *repudiate* the message (cannot say "I didn't get it."). Authentication systems may do *some* or *all* of the above.

Authentication Systems in Use Today

For the financial community, the American National Standards Institute (ANSI) specifies message integrity methods including check sums or message authentication codes. ANSI Standard X9.9 is the Financial Institution Message Authentication Standard in use today.

Importantly, note that authentication still allows the message to be read. For many applications, such as bank transfers, giving someone knowledge of the transfer and even its amount is unimportant, but authenticity, integrity, and nonrepudiation are absolutely essential.

The most current authentication method is the RSA technique, which is widely used, though not a U.S. government standard. In contrast, in 1994 the federal government approved its own standard developed by NSA called the Digital Signature Standard. Now there are two standards to deal with.

The RSA Public-Key Technique

The RSA approach uses the principle that if two large prime numbers (e for encryption and d for decryption) are multiplied together, the resulting number n is then very hard to factor back into the two original numbers (see Fig. 4.3).

In arithmetic terms:

$$n = ed$$

A message sent using n as the key can only be decrypted by a receiver possessing e or d. Since e never leaves the transmitter and d never leaves the receiver, and factoring n into e or d is not possible in any reasonable time, the system is secure. Since n appears in public, it is called the *public key*. Key d is a *private key* that never leaves the receiver; e never leaves the sender.

RSA is referred to as an *asymmetric* cryptosystem: it is easy to do, but very hard to undo. Used in another sense but equally true, the decryption key d is different from the encryption key e. Since

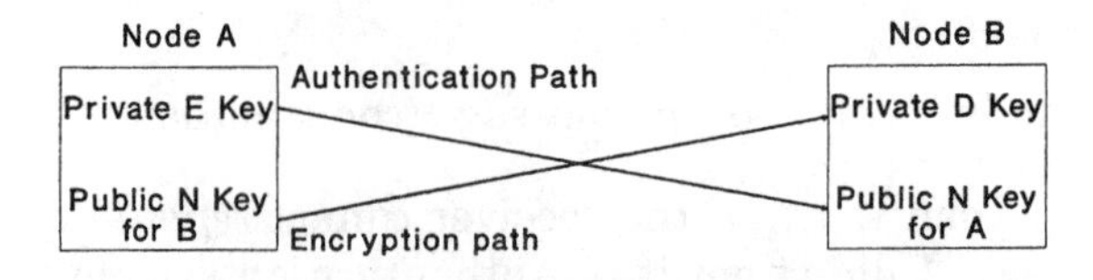

Encryption:
Node A uses Node B's Public Key to send a secret message.
Only Node B can factor Public Key B using its Private D Key.

Authentication:
A message authenticated using E Key in Node A is sent to B.
Node B uses A's Public Key to authenticate it.
Since E is secret, only Node A could have sent it.

Other stations with A's Public N key can read it,
but the point remains that only A sent it.

Figure 4.3 Two-key encryption and authentication.

either e or d if revealed can unlock the system, RSA is referred to as a *two-key* system.

Since n can be released safely to the public, it is practical to build a public-key directory that vastly simplifies the major headache of encryption, key distribution. For a 1,000-node network, 1,000 pairs of keys are needed. Using private keys the figure is much larger and is computed by the formula

$$\frac{q(q-1)}{2}$$

where q is the number of nodes. Thus the number of nodes is multiplied by the number of nodes less 1 (since a node cannot talk to itself). The result is then divided by 2 since one node-to-node key is used by both connected nodes. Given 1,000 nodes, then, the number of keys required is 499,500, nearly 5,000 times the number of keys required in a public-key system.

The unique value of a public key n can be assigned to an individual or a company and so used for authentication. This is the purpose of a *certification authority*. As we shall see, one such authority already exists and another is being formed.

When will computational power reach the level needed to factor n? A Cray 1S supercomputer in the mid-1980s was able to factor a 70+ digit number in under ten hours. In 1988, a 100-digit number was factored using a network of 50 small computers. In April 1994, a team from Bell Communications Research (Bellcore) was able to factor a 129-digit number in about eight months. Their technique used the Internet to distribute pieces of the problem to over 1,600 computers, a form of massively parallel processing. Some 600 volunteers in 24 countries permitted their computers' spare capacity to be used for the purpose (a "good" virus?).

Ronald Rivest, one of RSA's trio of developers, was on the Bellcore team. Afterward, he said, "This does not break the RSA system...but it does help us benchmark the system. One certainly needs to use more than 129 digits in a key." RSA Data Security, Inc., of which he is a founder, recommends that companies use key lengths of between 150 and 230 digits.

For example, RSA numbers for *e* and *d* are *each* 154 digits or 512 bits in commercial implementations. The public-key number for *n* in that case is *six digits,* beyond present capabilities. However, the Bellcore group stated at the time that it was "dangerously close" to cracking 155-digit keys.

Two-key authentication

Let us say that the sender "encrypts" a message using its secret *e* key. The receiver "decrypts" the message by using the sender's *n* key. Since no one else but the sender had the *e* key, the message came from the sender for sure.

The message is not encrypted in the sense of trying to make it secure, because the public *n* key can be used to make it readable. That is, any station can use *n* to decipher. The point is that *e* was in only one possible place, and thus authentication takes place.

In RSA, both authenticity and secrecy can be combined. The authentication process as above is used first. Then the authenticated message is encrypted using the recipient's public *n* key. The receiver uses its *n* and *d* keys to decrypt first. Then using the sender's *n* key, it authenticates the message.

Public Key as a Digital Signature

Public-key cryptosystems are easily adapted to add signature capability to authentication. This is a way of saying, "You told me this XXXX, and here is proof that it was you." Using a digital signature is a way of preventing *repudiation* ("I never told you XXXX") by the sender.

When the receiver authenticates the message, it stores the encrypted text, the decrypted plaintext, and the sender's public key. Since only the sender has the *e* key, and only the recipient has the d key, there is a firm logical connection between sender and receiver, making repudiation by the sender impossible.

Such a system does not protect against the reverse: "I told you XXXX ." While such a message if sent would be recorded, the lack of such a record at the receiver could mean that the message did not get through or that it was improperly erased. To provide such protection, an acknowledgment to the sender by the receiver must be used. Thus signatures must be used both ways for complete signature authentication.

Messages passed from station to station may have their signatures removed, the message changed and a new signature added. Consequently, the signature on a message authenticates only the *last* person (or system) to send it.

Allocating Public-Key Signatures

As public keys come into greater use, users will be given public-key signatures embedded in electronic certificates. Banks, notably Bankers Trust and other private organizations such as National Semiconductor and Trusted Information Systems, Inc. (T.I.S.), are gearing up to become certificate authorities (CAs) in anticipation of a $50 billion market by the year 2005. Lost, stolen, canceled, or fraudulent signatures will be reported to a CA so that all other CAs can be so apprised. Escrowed keys have also been found useful as a backup mechanism should a keyholder become ill.

As an example, the Internet Society established a CA system for its Privacy Enhanced Mail encryption standard. The federal government has set up a Certification Working Group consisting of representatives from the Department of the Treasury, NIST, the Department of Justice, NSA, and the FBI to work toward establishing a CA for the Digital Signature Standard. Potentially, there could be millions of keys held by the public and business to communicate with the Internal Revenue Service alone.

The American Bar Association plans to issue "Rules of Practice" for certificate authorities to cover the legal and procedural aspects of authentication. RSA Data Security, Inc., is expected to be at the top of the issuance chain. Readers should also see Benjamin Wright's books, *Legal Identity and Signatures on the Information Highway* and *The Law of Electronic Commerce,* both published by Little, Brown (New York, 1994, 1995). The number of CAs is expected to be strictly controlled to avoid the "Certificates-R-Us" problem of having too many CAs.

This is already beginning to happen. Bankers Trust and T.I.S. have proposed separate schemes to the federal government. The government has set up its Federal Information Security Infrastructure Service Program management office to define security policy and set technical guidelines. An examination of Commercial Automated Key Escrow, or CAKE, is part of their charter.

What Authentication Can and Cannot Do

An authenticator in its simplest form only validates who sent the message. It does not hide the message, nor does it say that the message as received is the way it was sent. The message could have been accidentally changed (as by a line error) or deliberately corrupted.

To add data integrity, the message may be encrypted. Then an adversary will not know what he or she is changing. A line error will cause decryption to fail, causing the receiver to ask for retransmission.

Another way to add data integrity is by a technique called *hashing*.

Hashing Algorithms

A form of authentication that provides data integrity but not send or receive authentication is called *hashing* (see Fig. 4.4). The algorithm manipulates ("hashes") the data in a carefully defined way and adds a signature reflecting the result. Popular hashing algorithms include Kaliski's MD2 algorithm (1992), Rivest's MD5 algorithm (1992), and NIST's Secure Hashing Algorithm (SHA). SHA is considered to be the most secure at the time of this writing.

The receiver performs the same computation. It compares its own result to a table. If the results match, the data have not been changed between sender and receiver. The CRC check to detect viruses is a simple hashing algorithm based on division of the data by a polynomial. The result, a remainder, becomes the CRC signature.

Comparing the computed CRC to a table reference is necessary. Otherwise, the data could be modified along with the CRC, resulting in a good check despite corrupted data. One way to minimize this possibility is to bury or distribute the signature in the file somewhere, making it harder for an adversary to find.

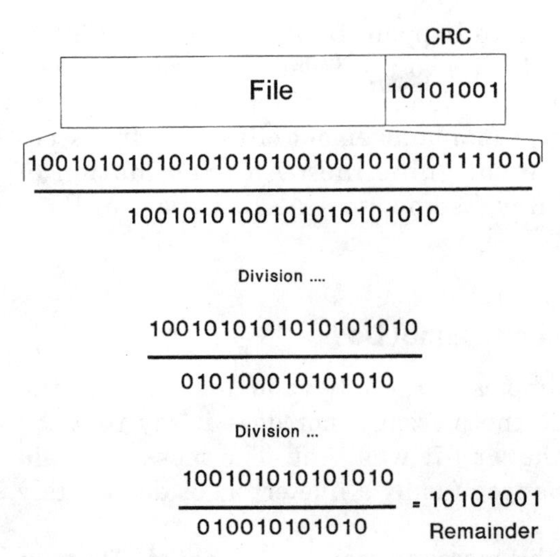

Figure 4.4 Hashing.

A better way is to encrypt the data plus signature or at least the signature. While a decryption key would be needed, a table of software signatures would not. A variation is to encrypt the signature using a public key. The private key would recover the signature and then use it for verification.

Hashing applications

One way to use hashing would be to add a separate signature, called a *file seal,* to published software. Virus checkers would have the signatures and compute them using the published software. If the file seal and virus checker signatures do not match, corruption of the file has likely occurred. The signature would not have to appear in the published software, only as an encrypted reference in the antivirus software. For that matter, as a utility, the same computation could be run for any desired file if it is known to be good beforehand.

Possible complications

A problem with any hashing algorithm is that some files are legitimately modified to insert user information such as a name or customization options. This variable information would have to be in a separate file or a portion of the file excluded from the signature computation. Such information would remain vulnerable to a virus or tampering.

Another Public Key: The Digital Signature Standard

For years, NIST promised the user community to come up with a public-key standard, just as they did for private-key DES in 1977. With DES's adoption, the use of private-key encryption was legitimized. It was hoped that the public key would achieve the same respectability for reasons of legal liability, compatibility with other users, and especially simplification of key management.

It was widely hoped that NIST would adopt the de facto standard, RSA. But NIST did not: they instead proposed their own creation called the Digital Signature Standard (DSS). A comparison of RSA, DSS, and PEM (Privacy Enhanced Mail) is shown in Table 4.1.

The reasons for NIST's choice were threefold. First, RSA wanted royalties for its technology whereas NIST sought a public domain system. Second, NIST desired something exportable to other countries. Third, RSA was privately conceived and not developed by NSA. It is widely believed that RSA was rejected in part due to the "not invented here" syndrome.

Public Key Partners, Inc. (PKP), which holds the patents used by RSA Data Security Inc., then claimed that DSS used its technology. After a two-year battle between NIST and PKP, NIST finally gave up. NIST announced that users (vendors and private users, but not government users) would pay royalties to RSA for DSS technology. Export controls would also apply to DSS.

This decision met with a storm of protest from over 100 members of the user community including IBM, AT&T, Citibank, N.A., and the Bell operating companies. It was also opposed by the Canadian government and the European Community. Their collective objections brought the NIST–RSA controversy back to the beginning.

In May 1994, under intense pressure from the Executive Branch, the Department of Commerce dismissed PKP's patent claims (PKP's patents are used by RSA Data Security, Inc.) and declared DSS openly available and a legal substitute for a handwritten signature. DSS has accordingly been designated FIPS 186 and at the end of 1994 became a purchasing requisite for government systems employing EDI, e-mail, and other uses.

There are a considerable number of planned DSS applications:

- The military Defense Message System, consisting of over two million users.
- At least six Internal Revenue Service applications ranging from wage reporting to access by 200 million individuals to tax infor-

TABLE 4.1 Comparing RSA, DSS, and PEM

RSA	DSA	Internet PEM
Well known; over 10 years' experience	No experience yet	Same
A de facto standard	Becoming a federal standard	RFC
Used by many financial institutions	Designed for IRS and the Treasury Department	
A de jure standard (ITU-TSS X.509; ISO 9796)	A de jure standard	
Not federally sanctioned	Federally sanctioned	
Slower signature generation (50 ms)	Faster generation (40 ms)	
Faster verification (3 ms)	Slower verification (80 ms)	
Sign and verification times for a 10-k message not specified	Sign (10 kbyte message): 1.5 s Verification (same): under 3 s	
Encryption capability	Signatures only (encryption to be announced)	DES
Privately developed	Created by NSA	Created by IETF
Proprietary	SHA hashing	MD2/5
RSA	DSA signature	RSA
License fees	License fees	
Export controls	Export controls (see FIPS 180)	

mation and electronic delivery of tax forms. Implementation of these systems is expected by the year 2000.

- Use of DSS by the Federal Aviation Administration to issue pilots' medical certificates.
- Use by NASA for network access and for funds control, authorizing millions in funding for programs from the Office of Management and Budget down to fine levels of granularity such as individual projects.

This does not mean the legal wrangling is over. The federal government said that it would defend vendors selling DSS products to the government against patent suits, but it would not necessarily do so for suits brought against a vendor as a result of commercial sales. However, several vendors, including AT&T, have gone ahead with DSS implementations without obtaining a license from anyone.

PKP may not bring lawsuits against the government directly, but against smaller companies working on government contracts. Under present policy, the government would defend such companies at its expense in court.

There are indications that the government is continuing to negotiate with PKP behind the scenes. For those in the information processing field, and especially those involved in electronic data interchange or any aspect of financial communications, this ongoing dispute merits attention as FIPS standards tend to become industry standards as well.

Aside from the legal problems, there are several technical snags. To begin with, DSS provides only for document authentication, not encryption. Compared to RSA, DSS is therefore half a standard. Furthermore, NSA, working through NIST, withdrew the Secure Hashing Algorithm portion of DSS when NSA discovered flaws in the SHA technique. A revised SHA became effective in October 1995 as part of the Secure Hash Standard (FIPS PUB 180-1).

The end user community may get the worst of all worlds, starting perhaps with royalty payments for incomplete technology that today exists partly on paper and ending with a confusing dual standard. Had the government simply adopted RSA two years ago, users could have been two years' farther down the road and had a complete single standard that was already in wide use and legitimized in ITU-TSS X.509 and ISO 9796 standards.

Part of the motivation for developing its own standard is that NIST was pressed by NSA into backing DSS. Perhaps NSA and NIST thought that they could work around patent infringement claims. The notion of paying a small company millions in royalties surely did not sit well with the government or with the vendor community who would actually pay them. Lastly, NSA certainly remains concerned about its ability to conduct surveillance when RSA is used.

The Digital Signature Algorithm and key management system on which DSS is built is also the heart of another hardware-based public-key cryptosystem titled *Capstone*. Capstone is to be used by the military and civilian communities. The patent battles have held Capstone up, too.

The ANSI X9.F.1 Security Committee Working Group submitted a proposed public-key standard to ANSI incorporating both RSA and DSS. This dual standard, intended to appease the financial community, is sure to confuse everyone. Said one user, "It's just like old times."

As a possible compromise, IBM suggested that NIST adopt ISO Standard 9796 (which includes RSA) as an alternative to DSS. NIST said at the time that it was open to this possibility, but has taken no action.

Perhaps the most sensible reaction was that of CSSPAB, who recommended approval of DSS by the Secretary of Commerce only after a national public review of cryptography policy. DSS, however, was approved under pressure from Vice President Al Gore's office, who regards DSS as essential to the National Information Infrastructure.

What is known about DSS

In the midst of all the controversy and secrecy, some nuts-and-bolts facts are known. To clear up some terminology, the *Digital Signature Algorithm* is specified in the *Digital Signature Standard*. DSS provides only authentication and data integrity. It provides no encryption.

There are three parts to the DSS:

- The Digital Signature Algorithm (DSA creates signatures)
- The Secure Hash Algorithm (SHA provides data integrity)
- The key management element (as yet unspecified)

The Secure Hash Algorithm

As Fig. 4.5 shows, a message of less than 2^{64} bits is entered into the Secure Hashing Algorithm (SHA). The hashed result is a 160-bit output called a *message digest*. A private key is then added to be

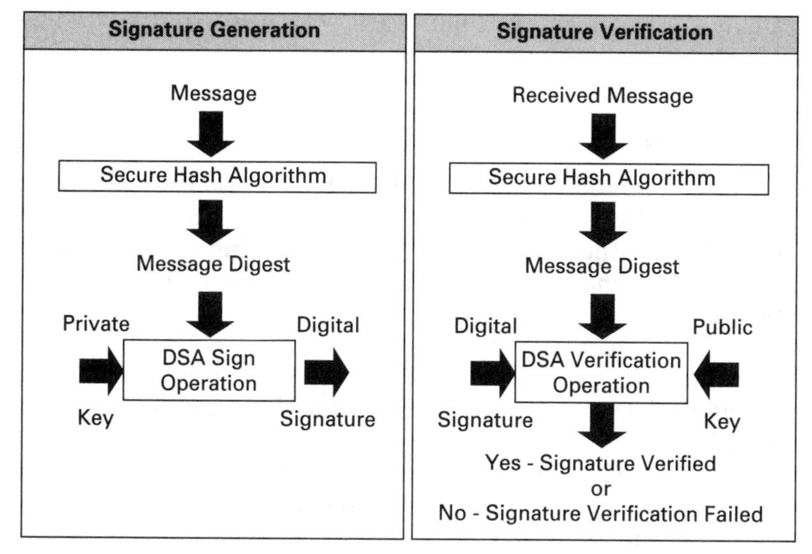

Figure 4.5 The Digital Signature Algorithm. (*Source: FIPS Pub. 180.*)

used as a signature. Between DSA and SHA, an authenticated message of known integrity is created.

At the receiver, SHA is run on the message, again creating a message digest. If the hash result is the same and the signature is verified (using a public key in the receiver), the message is authenticated as from the sender and has data integrity.

Any change to the message itself has a very high probability of changing the message digest, resulting in a failure to verify the hash signature. Posed another way, it is "computationally infeasible to find a message which corresponds to a given message digest or to find two different messages which produce the same message digest" (from FIPS 180). Signing the message digest rather than the message itself often improves efficiency because the digest is usually of smaller size than the message.

Uses for SHA

Some applications for SHA include:

- Electronic mail
- Electronic funds transfer
- Software distribution
- Data storage

SHA may be implemented in hardware, software, firmware, or any combination.

Digital Signature Encryption Using RSA

In the RSA technique, a hashing algorithm functionally similar to SHA creates a hashed "profile" of the document. The profile is then encrypted using the sender's private key, creating the document's signature. This encrypted signature is attached to the document for transmission.

At the receiver, the recipient uses the sender's public key to decrypt the signature. It compares the decrypted profile to its own computation. An identical profile means the document has not been tampered with.

RSA's technique can be used for document encryption as well as creating digital signatures. DSS creates only signatures. If DSS and RSA appear to be very similar, that is why there was or is a patent fight.

Authentication via Dial-Back Systems

Authentication takes the form of dial-back in hardware systems (see Fig. 4.6). Let us say a user wishes to connect to a computer via dial-up. The user first dials a number. The call is answered not by a modem but by an electronic barrier—the dial-back system. There is no telltale answering modem tone. The caller enters a code on the pushbutton phone and hangs up.

The dial-back system consults its table and finds the caller's phone number associated with the code. The system calls back and the connection takes place. The dial-back system has authenticated the user's location and the user him- or herself.

So what can go wrong?

First, it is easy to record pushbutton-phone access tones. Machines exist that can do it quite easily. These machines are so simple they can be built at home.

When the system calls back, it is really authenticating a *place*, not a *person*. In addition, this involves two calls per session resulting in a bigger phone bill and wasted user time. What if a thief records the access tones and then places the callback phone on call forward? The thief could divert the call to anywhere. Hackers use a technique called *glaring* to outwit a callback system that cannot detect a dial tone. They just stay on the line after feeding in legiti-

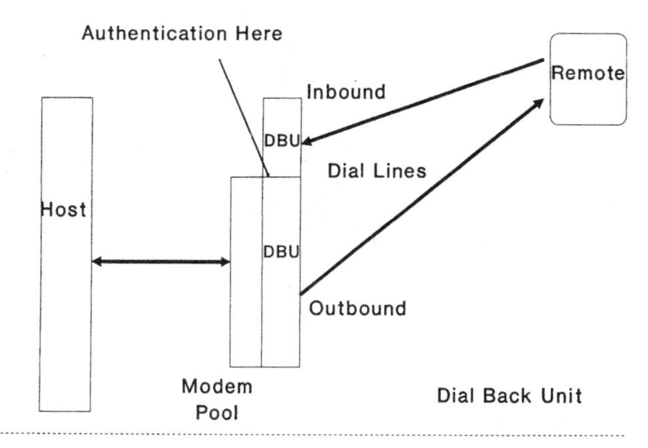

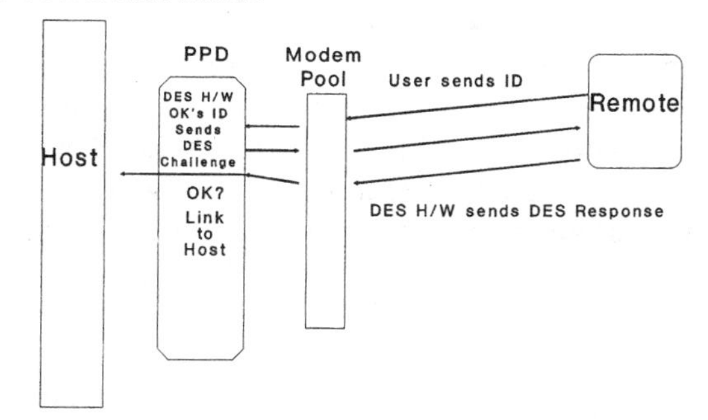

Figure 4.6 Dial-back unit and PPD systems.

mate tones. The callback unit dutifully pulses out dial tones but to an already open line where they have no effect. Once the dial sequence is over, the cracker can start working on your internal passwords. In the view of James Settle, supervisor of the FBI computer crimes squad, "Forget callback. You might as well open the door and let them in."

The above method is a *fixed callback system*. It is not only relatively insecure, but it is inflexible. If the caller moves around a lot or is at another location, the system will not work unless the callback number is known to it. One way to add some flexibility is to create a code for every location where a user might be. This gets to be very bulky and still does not address the traveling road warrior.

Some systems support both fixed and *variable callback*. Variable callback is more flexible. It will call a specified number back once a proper code has been entered. This is not particularly secure if the same pushbutton code is used over and over.

The manner of code entry is important to validate the user. Pushbutton tones can be recorded or decoded easily. A better choice is a hand-held calculator "smart card" device (e.g., SecurID by Security Dynamics of Boston, Mass.), where the code changes every minute or so.

With a smart card, the authentication aspect of the callback becomes unimportant. Usually, the desirability of central billing and audit control is the driving force behind a variable callback system. Smart cards are far more reliable authentication devices than a simple callback.

Port Protection Devices

Port protection devices (PPDs) provide much greater protection than a simple dial-back scheme. PPDs do not depend on a callback. Instead, they use a DES challenge–response system. When a user begins a call, they enter a password into software located in the workstation. The workstation software sends this identifier via modem to the PPD that validates the ID and sends the workstation a DES-encrypted one-time password challenge. The workstation software answers the challenge and the user may or may not be connected.

PPDs, being digital, are not limited like callback systems to 12 pushbutton tones for password combinations. The PPD has a much larger domain of possible passwords. If there is no workstation, the same sequence may be performed by a user with a smart card and a dumb terminal.

Sometimes the challenge–response software will work directly from an application (such as Digital Pathways' developer's kit) or from communications software (Racal-Guardata). Since the entire exchange is encrypted, recording the sequence is worthless. Some systems (such as Racal-Guardata's) even encrypt the entire session.

Protection systems

At the host end, the protection system may sit between the telephone network and the modem(s) or may sit between the modem(s) and the communications controller or server. The general preference is for the former, as it is easier to implement *service masking*.

Service masking means that the protection system answers the call, not the modem. As a result, there is no modem tone to tell callers that they have found a modem port. Hackers with demon dialers depend on detecting modem carrier signal tones before they start entering possible passwords. Some systems answer with a synthesized voice and recognize caller voice patterns. This is excellent (biometric) authentication.

Modems and software installed at remote sites should be internal to the workstation; otherwise it is too easy to bypass the entire system. For instance, some modems for remote sites contain encrypted, embedded passwords that are exchanged between modems (Microcom Corp., Norwood, Mass.). Such systems should also involve a password entered by the user. Otherwise, stealing the remote equipment will gain the thief access. This same principle applies to notebook computers with login scripts containing passwords.

Group management lets calls be returned on different lines than those that made the calls. This permits a small number of answering lines and a larger number of callback lines.

You have probably had the experience of picking up the phone to make a call only to hear someone on the other end trying to reach you. The same thing can happen in dial-up operations. As the computer calls the user back, another inbound call could seize the line and be connected.

There are two solutions. One is to give out only the inbound numbers and keep the dial-outs private. Privacy reduces the risk. This does not stop the hacker's demon dialer. To stop these hackers, the outbound lines can be disabled from accepting inbound calls if run through a PBX or Centrex service. Some PPDs' logic can be set to go outbound only.

Group management may also be used to elect variable callback for the sales force and fixed callback for less mobile callers.

All systems should be *scalable:* one should be able to start small and work up in size easily. No system should require any changes to the host machine.

Management features

All PPD systems should have an extensive *audit trail* capability. At a minimum, the audit trail should show:

- Unsuccessful and successful call attempts
- Access durations
- System utilization data
- Telephone toll charge data (e.g., calls by time of day)

Audit trails should permit user account password management. Although not mandatory, it is often convenient to have remote diagnostics, management, and configuration capability.

All systems should have a series of timers that terminate a call setup in progress if too much time goes by. Similarly, the manager may wish to lock out all callers during specified times, such as from 7 P.M. on Friday through 7 A.M. on Monday.

Single-port systems cost about $500 per end and usually involve two security modems or a sentry device at the host end and a DES-authentication device at the remote. Multiple-port systems carry a base price of about $4,000 at the host end plus $400 per port (including the remote end).

New features

One security modem vendor (Phonetics' ModemMate I; Aston, Penn.) offers a modem with a built-in voice synthesizer. Upon answering, it says, "Hello." Demon dialers hear the so-called voice and hang up.

Another intriguing approach is to use automatic number identification (ANI). When the incoming call rings, the ANI system gives the called party the caller's phone number. The PPD (or a modem such as Phonetics' ModemMate II) looks in its database for a match. Finding none, it terminates or does not answer the call. This is a far better approach than the crude dial-back method. The ANI approach is implemented by Quality Telecommunications, Inc., in its Compusafe product. Some security modem vendors also offer it.

Summary

In this chapter we have learned what is meant by authentication and how it is applied in electronic systems. Because it is less processing-intensive, authentication is popular when the contents of the message need not be hidden. For instance, it finds application in many banking and transaction-driven systems where secrecy is not necessary but the validity of sender, receiver, and message is.

In the next chapter we discuss the use of encryption and compare it to authentication. These two chapters together are intended to give you a firm perspective on the techniques we use to protect data, which is fast becoming organizations' most precious asset.

Review

- Authentication is the process of *determining accuracy.*
- We authenticate a sender's identity (*data origin authentication*).
- We may authenticate a sender's message (*data integrity*).
- We may authenticate a receiver's identity (*peer entity authentication*).
- Authentication is accomplished via something users know, have, or are.
- Added to a message, electronic *signatures* are the authentication mechanism.
- The RSA dual-key system is in widest use today.
- The RSA system makes use of the fact that two large prime numbers are hard to factor.
- The RSA technique may be used for both authentication and encryption.
- Authentication is more expensive than encryption, but faster than encryption if both processes are implemented in software. Hardware implementations favor encryption.
- Another technique, the Digital Signature Standard only authenticates.
- DSS, proposed by NIST, has been designated FIPS 186, a federal standard.
- *Hashing* creates a unique electronic profile of a document.
- The Secure Hash Algorithm is part of the Digital Signature Algorithm.
- Dial-back systems are in wide use but have several weaknesses.
- Fixed dial-back systems call a specific number, not a person.
- More recent port protection devices use encrypted authentication.
- New techniques such as using automatic number identification are becoming popular and are less bulky to use.
- Authentication is a subset of encryption, since encryption inherently includes basic authentication.

Appendix. Vendors

This is a partial listing.

Dial-back and port protection devices

Digital Pathways
Mountain View, Calif.
Defender

LeeMah Data Security Corp.
Hayward, Calif.
TraqNet

Racal-Guardata
Sunrise, Fla.
Access Gateway and Security Gateway

Port protection devices

Cettlan Corp.
Security Modems

IC Engineering, Inc.
Modem Security Enforcer

Microcom, Inc.
Norwood, Mass.
High Density Management System

Millidyne, Inc.
Raleigh, N.C.
Sequel and Auditor

Octocom Systems, Inc.
Montage

Phonetics, Inc.
Aston, Pa.
Modem Mate I and II

Quality Telecommunications
Compusafe

Raxco, Inc.
DialBack (for DEC VAX/VMX)

UDS/Motorola
Huntsville, Ala.
GlobalView

U.S. Robotics
Skokie, Ill.
Communications Server 386 (dial-back)

Western DataCom Co., Inc.
Westlake, Ohio
Security Modems

For information regarding licensing of DSS and SHA algorithms, call NIST's FIPS office at (301) 975-2816.

Chapter

5

Encryption Methods

Objectives

At the end of this chapter you will be able to:

- Define what encryption means and does
- Explain the basic working of the Data Encryption Standard (DES)
- Discuss the advantages and disadvantages of DES
- Discuss how and why encryption is useful as a communications security tool
- Explain how a single-key system inherently authenticates
- Specify the issues that pertain when encryption is considered
- Compare and contrast encryption with authentication
- Explain end-to-end and node-by-node encryption
- Explain why double encryption might be used
- Understand the proposed Clipper and Capstone cryptosystems
- Relate recent government initiatives toward establishing a policy that protects individual privacy, allows secure electronic commerce but denies criminals the secrecy afforded by encryption
- Apply encryption in voice and data applications

Encryption Basics

The study and development of ways to hide information and then recover it by only the intended recipient are the basis for *cryptol-*

ogy, a word derived from two Greek words, *kryptos* (hidden) and *logos* (word). Cryptology includes *cryptography,* which is the study of ways to disguise information. Opposing cryptography is *cryptanalysis,* the adversarial deduction of intelligence from encrypted text. The process of encryption and decryption is carried out by a *cryptosystem.*

Cryptosystems are by no means new. The early Egyptians, Hebrews, Babylonians, and Assyrians all had cryptosystems. Spartans and Romans encrypted military communiques. Encryption is discussed in an Arab encyclopedia dating back to 1412.

The process is shown in Fig. 5.1. Information or *plaintext* is converted into *ciphertext,* a *cipher,* or a *cryptogram*; the latter three terms are synonymous. The process is called *encryption* or *encipherment.*

Almost always, an *encryption key* is used to allow changes to the ciphering process on the assumption that the longer a key is used to encrypt traffic, the more analysis can be conducted against the traffic to derive the encryption key. A new key forces an attacker to start all over again. Periodic key changes are designed to give cryptanalysts nervous breakdowns. Sometimes it works.

A *decryption key* is needed to reverse the encipherment process. In a *single-key system,* it is the same as the encryption key. Single-

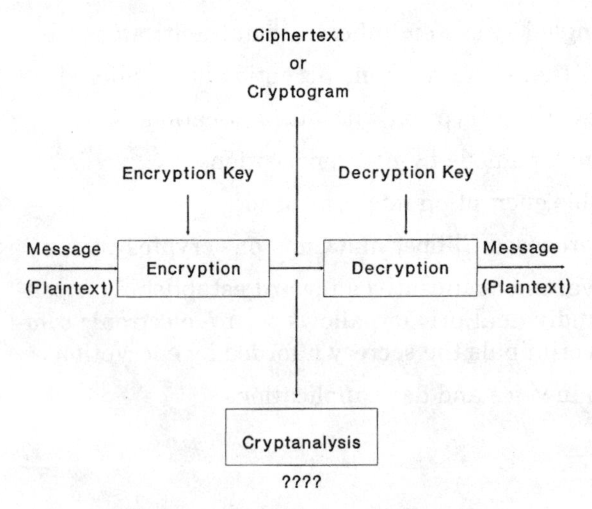

Figure 5.1 The encryption and decryption process.

key systems are also called *symmetric* cryptosystems. As we saw in Chapter 4, the authentication systems most often implemented today are two-key asymmetric systems.

The Data Encryption Standard

Encryption algorithms used to be a deep dark secret. This all changed in 1977 when the United States federal government adopted the Data Encryption Standard (DES).

Developed by IBM under contract to the National Bureau of Standards (now called the National Institute of Standards and Technology or NIST), the DES algorithm is publicly available to anyone who cares to look. The reason for this is that possession of the DES algorithm does not really help an attacker.

Finding the one right key from the *76 quadrillion* keys possible in DES is the real problem. To give you an idea of the difficulty, an attacker would have to test a billion keys a second in order to find the key in about a year.

The public nature of DES is a reflection of the old saw that "two can keep a secret if one of them is dead." With time, the odds of compromising a private algorithm increase. With a public algorithm, there is no such risk. Another benefit is that if public, the algorithm can be tested for weaknesses. As we noted earlier, DES is the most thoroughly tested algorithm ever, and no major weaknesses have been found since its debut in 1977.

DES implementations

For communications purposes, most DES implementations are in hardware encryption devices and use specialized DES chipsets. Software encryption and decryption are processor-intensive and press the resources of desktop computers. However, DES software implementations are commonly used to encrypt passwords and files in both network and non-network operating systems.

In 1994 the U.S. government reversed a long-standing policy that all networked DES implementations had to be in hardware. That policy was established on the basis that hardware implementations were easier to certify and harder for people to tamper with.

Conversely, software implementations were thought to be more readily compromised. The advantage to software implementation is that the product cost is generally lower and software is easier to install than encryption hardware with its attendant space and cabling needs.

The policy change has implications for private industry as well. In particular, the financial community will feel justified in implementing software encryption, pointing to the federal government as their model. It will be difficult for government auditors to disallow software encryption if the federal government is using it.

As one might expect, there is a price. NIST has published[1] a 56-page set of rules titled *Security Requirements for Cryptographic Modules* that establishes four security levels for cryptographic systems (see Table 5.1). These requirements are to be implemented in software security products that will then be tested with government-developed software by independent testing laboratories for compliance. *To protect yourself, you should obtain a copy of the certificate of compliance before buying or leasing any such software.*

Specific encryption algorithms are not named, giving the requirements more flexibility. Both public-key encryption and digital signature technology are incorporated in the document.

How DES works

Fundamentally, the DES key, 56 bits long, is used to generate a pseudorandom stream of 1s and 0s (see Fig. 5.2). The stream is not really random. It will begin to repeat one day. But the period is so long that it appears random for practical purposes. Furthermore, changing the key starts a new and different stream. This stream is mixed with data, and then the mixed stream is fed back into itself. The result is very scrambled data.

TABLE 5.1 Security Requirements for Cryptographic Modules*

Level	Description
1	Lowest security level; supports cryptographic functions in application software, integrated-circuit boards, and add-on equipment.
2	Requires C2-trusted operating system or equivalent for multiuser operating systems; tamper-resistant seals and locks on desktop computers required.
3	Requires B1-trusted operating system or equivalent for multiuser operating systems; secure and nonsecure data ports must be physically separated; must have the ability to authenticate a user's identity.
4	Highest security level; requires B2-trusted operating system or equivalent on multiuser operating systems. Level 4 systems can detect system penetration and then automatically erase stored data.

*These security levels apply to *unclassified* information residing on computers or traveling via telecommunications systems.

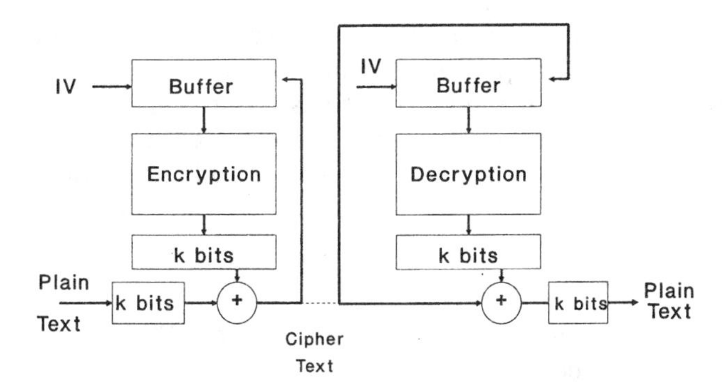

Figure 5.2 How DES works.

The decryption process is designed to generate the same pseudorandom pattern, starting at the same point as the sender. Exactly the reverse process takes place, subtracting by iterations, to arrive at plaintext data.

Given the above, it is necessary that both sender and receiver have the same key. This inherently authenticates one to the other. Even in a multipoint data network, where all stations have the same key, they still authenticate to one another, although each station's specific authentication identity is lost. This is another way of saying, "As long as you are a known member of my group, I do not care who you are."

Data integrity is preserved because an attacker has no idea of where to begin. Even if an attacker makes a known plaintext attack, say on a data line's synchronous data link control (SDLC) header, that SDLC header will be encrypted differently the next time and the hapless attacker has no way of knowing how.

DES keys

DES keys are 56 bits long, plus 8 bits for parity. The parity bits are not part of the key. If you multiply out 2^{56} possibilities, it comes out to 76 quadrillion (a little over 7×10^{16}) key possibilities.

People do not usually create 56-bit keys by hand when they use DES. The system they are using automatically picks a key. This is more convenient, is more secure, and allows *weak-key avoidance.*

DES has four known weak keys in which encrypting once and then again with the same key produces plaintext. Six other pairs of keys will produce mirror-image key extractions. Double encryption of these pairs with the same key will also produce plaintext. To prevent this, DES cryptosystems detect these keys and throw them out.

DES online

Used in a wide area network as shown in Fig. 5.3, a DES encryption device resides between the modem, the data service unit or channel service unit, and the terminal device, whether a workstation, host computer, multiplexer, or facsimile machine.

DES devices used this way are insensitive to bit format; they simply invert or do not invert bits in a row as they are presented. This creates a problem for devices such as T-1 multiplexers where framing bits must be recognized and 1s density requirements must

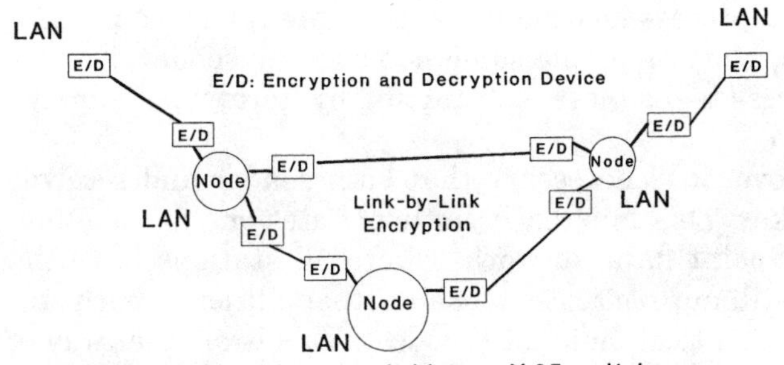

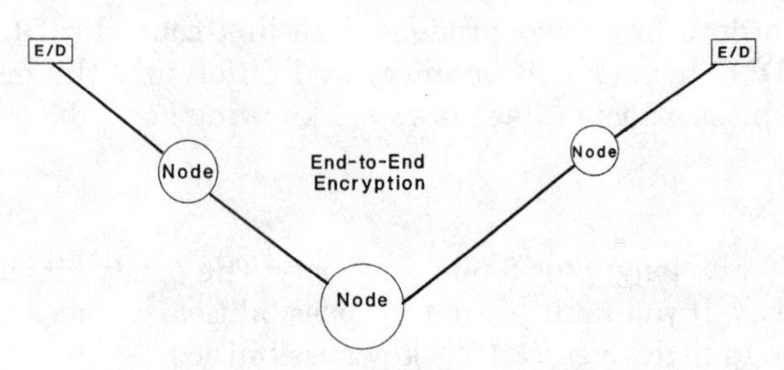

Figure 5.3 WAN and LAN internetwork encryption.

be met. For such cases, only the 192-bit data field for T-1, for example, is encrypted. Sufficient 1s are inserted to ensure that the observed density is three 1-bits in the 24 bits sent. Similar rules are applied at T-3 rates.

Line errors will force the two DES devices out of synchronization. If DES's single-bit cipher feedback mode is used, the machines will fall back into synchronization after 63 bit times following a 1-bit error. Using an SDLC or binary synchronous protocol, it will appear to the system as a defective frame from which the system will self-recover. For this reason, physical layer encryption systems are very transparent to the systems they protect.

DES encryption devices are available from the slowest asynchronous speeds up to T-3 (44.736 Mbit/s) rates and higher. Processing delay depends on the encipherment option used, but is negligible (e.g., 1-bit time) compared to other system delays.

Wide Area Network Techniques

Encryption can be performed link by link as shown in Fig. 5.3. This is sometimes necessary if the nodal processors need to read plaintext addresses or control information, as in X.25 or TCP/IP networks. However, the traffic will be in plaintext while in the switch or router, causing a potential security weakness. Such an Achilles' heel may not be acceptable.

An alternative is to use a more intelligent encryption device that encrypts ISO transport layer information and higher, while leaving the routing and node-to-node information in the clear. The end-to-end information is thereby protected, while leaving necessary control information in plaintext. The disadvantage is that the source-to-destination data may be sensitive enough that even knowing its Internet protocol (IP) address, for instance, may yield a real clue as to the traffic's source and destination(s). Encryption techniques for TCP/IP systems are discussed in Chapter 10.

End-to-end encryption

It is certainly more secure to encrypt at one end, transparently carry the traffic, and then decrypt at the other end. In this way, data do not exist in plaintext form at intermediate nodes. It also saves tremendously on encryption devices and greatly simplifies key management.

Some encryption devices will, through the random process, generate control codes that will be acted on by an intelligent network. Most encryption devices encrypt everything, so an escape sequence is needed in an intelligent network telling the network to ignore what appears to be control information until the escape sequence is again presented. When a device mistakes one condition for another, as in mistaking encrypted data for control information, this is called *aliasing*.

The best way to encrypt through an intelligent network is to encrypt the traffic before it enters the network. The network can thus create a control packet around the encrypted traffic. The network must not react for control purposes to anything in the packet's information field. This method does not require escape characters as the information field by definition cannot contain control information. Accordingly, the intelligent network should not look for any. Not all systems, meaning some X.25 packet switching networks in particular, can support end-to-end transparent connections.

Multiple encryption

For very high security applications, it is not unusual to encrypt once and then encrypt the enciphered stream again with a different key. Even if one key is stolen or recovered by cryptanalysis, the adversary still has to analyze and break the other one.

Traffic security

The simple absence or presence of traffic on a network gives an adversary clues as to work habits, a plan in the process of execution, and so on. To prevent traffic analysis, many cryptosystems send strings of bits all the time, whether there is traffic or not. If there is no traffic, there is simply no output to the user from the cryptosystem. Accordingly, end point systems are unaffected.

The use of traffic security depends on the medium employed. A truly shared medium such as a LAN cable cannot use it or else every station would be talking all the time. The exception is where an intelligent concentrator is used. Such a concentrator is really a multiport bridge, which has the effect of isolating each workstation. No longer a literal party line, traffic security can be invoked without fear of upsetting the media access method.

LAN Network Techniques

LAN protocols have control information at their beginning, such as address information. Most often, encryption is only performed on the information field as was illustrated in the IEEE SILS frame of Chapter 3.

Frame encryption is usually done in software. Frame encryption is not too common in LANs because the workstation or server processing overhead is very significant.

It is possible to encrypt the entire frame except for the preamble, which must remain clear to establish timing. The preamble carries no information. This way, the identity of destinations and sources can be hidden as well as information carried back and forth.

LAN internetworks can use hardware encryption between routers and bridges. Frames can be bridged or routed if their medium access control (MAC) and routing information is not scrambled. Since routing information is often found in the information field, care must be taken when encrypting routable traffic. Generally, it is just easier to use hardware link-by-link encryption for absolute security and transparency.

Encryption Key Management

Key management is the bugaboo of encrypted enterprise networks. Like any asset, keys must be generated, distributed, and accounted for. This becomes difficult in a multinational network. The likelihood of a failure somewhere must also be recognized.

In a single-key system, there must be one key per point-to-point link, shared between either end. Changing the key must be done manually at a set interval. The interval is usually 24 hours, but may safely be extended to a few days. There is no absolute rule because the sensitivity of the data is the prime determinant. In the most basic systems, keys must not only be changed manually, they must be distributed manually. The entire process is fraught with expense and opportunity for error.

ANSI Standard X9.17 defines procedures for manual and automated management of keying materials, and uses DES for key management in a variety of operational settings. FIPS PUB 171 specifies a set of X9.17 options for federal government users using X9.17 protocols.

Due to the expense and labor-intensive nature of key distribution, it is accepted practice today to distribute new keys via an encrypted network directly into the encryption devices that will use them.

Another option is to use public keys. Since the public key for any station need not be secure, a public-key directory allows any station to send a message to a station simply by asking the directory for its public key. Public keys recast the keying problem from unmanageable to manageable. As we saw in Chapter 4, for a 1,000-node network, the reduction is from 499,500 keys to 1,000 public-key pairs.

The Public-Key Infrastructure

Using a public key requires a way to connect the identity of a user to his or her public-key signature logically. This is easy. The hard part is certification that the linkage is still valid. If a key is stolen, or if a person changes jobs, the linkage must be broken.

A study completed by the M.I.T. Research and Engineering Corp. for NIST placed startup public-key costs at over $1 billion. That cost is for the government alone, excluding the private sector. Maintenance costs were estimated at over $2 billion a year.[2] Using the X.500 Directory Service as a base technology, MITRE speculated that the Federal Reserve Board, the U.S. Postal Service, the General Services Administration, or a private contractor could handle the task.

There are already several certification authorities in existence. RSA Data Security, Inc., Trusted Information Systems, Inc., and the Massachusetts Institute of Technology are three examples, but none of them operate on a scale sufficient for interenterprise applications.

Furthermore, the liability concerns are significant. Only the government can handle the risk on the scale that is envisioned. While on the one hand the government can insulate itself from lawsuits at will, business will be chary of using the service if the government will not accept any responsibility for its actions.[3] At this writing, the Department of Justice refuses to accept this responsibility.

Platforms for Encryption

DES is designed to run in hardware. Over 100,000 DES chips are sold annually. Attainable hardware speeds are very high—over 40 Mbit/s. Cost can range from a few hundred to tens of thousands of dollars. Speed, key management capabilities, flexibility, and physical security features all determine cost.

Software implementations are available, but run much slower than hardware. There may be a software cost advantage in that software DES can be duplicated easily at little expense, as might be used in workstations. Of course, this depends on the licensing terms for the software.

Many network operating systems use software DES as a utility. NetWare (all versions) encrypts password keys on the cable from workstation to server and at the server. NetWare 4.X uses RSA public-key security, which is available for NetWare 3.11 as an NLM called NetWare Care Protocol (NCP) Packet Signature. NetWare 3.12 includes this NLM. Lotus' Notes and Microsoft's Windows for Workgroups also use RSA public-key security.

Novell's UnixWare workstation operating system offers encryption utilities. The utilities allow DES encryption and decryption at the kernel, library, and command-support levels. Files are encrypted according to a user password. When the correct password is supplied, the file is decrypted. It is encrypted again when the user is done.

Banyan VINES encrypts *everything* above ISO layer 3 on the LAN cable.

Workstations may also use DES in hardware or software form to handshake with a port protection device guarding a PBX or a protocol converter. Workstations may also use DES as a file transfer or interactive utility. File encryption products using DES (e.g., RSA Data Security's Mail Safe) are quite common.

A word of caution

Archiving programs such as PK.ZIP or dynamic compression–decompression programs such as Stacker will not compress encrypted files effectively. Archival programs seek repetitive bit strings to reduce to smaller string symbols. Encryption is specifically designed to avoid such repetitive strings, and so compression is ineffective. Instead, reverse the order: PK.ZIP the file first to make it smaller. *Then* encrypt.

DES Single Key versus (RSA) Public Key

DES is faster than RSA in real-time applications. Public-key systems need to sign a message and verify it at the other end; this takes longer. The ideal combination is to use RSA to distribute DES keys and then use the faster DES online (Fig. 5.4). DES requires more keys whereas public keys require fewer ones. Key management with a public-key system is much more workable.

Since DES keys must be the same at both ends, authentication is assured intrinsically as noted before; the encryption is almost a bonus. Public-key systems may do authentication, encryption, or both.

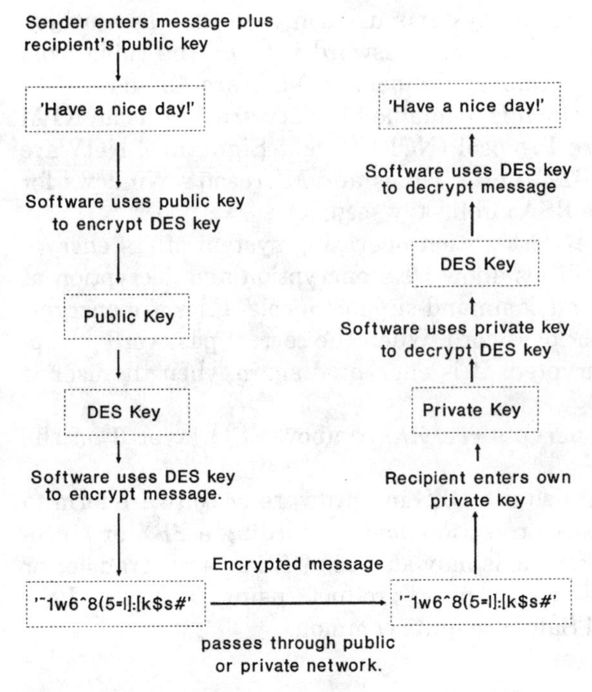

Figure 5.4 DES and public key together.

DES equipment is generally less expensive than the more processing-intensive public-key systems. For a comparison of encryption and authentication, see Table 5.2.

Voice Encryption

DES and public-key systems may be used to encrypt data, facsimile, or voice, but DES is preferred because it is faster. Indeed, some facsimile machines come with a built-in encryption device or an EIA-232 Data Terminal Equipment (DTE) port on them for an encryptor.

There are several voice encryption products on the market. One of them is the STU-3 telephone made by Motorola. These have been made to NSA specifications for many years. A newer device is the AT&T Secure Telephone Device incorporating the Clipper chip. Some 8,000 of these devices were ordered by the FBI. A variety of proprietary voice scramblers are sold using frequency-swapping techniques (for example, exchanging a high frequency for a low one, using a pseudorandom sequencing method).

TABLE 5.2 Comparing Encryption and Authentication

Characteristic	Encryption	Authentication
Number of keys	2_{56} (DES); 2_{80} (Skipjack)	156 digits (typical) $\times$ 2 (for RSA)
Authenticate originator?	Yes: whether single key or two keys, the keys must match	Yes: this is the purpose of authentication
Authenticate information?	Yes: it cannot be tampered with if it cannot be identified	No: not in basic authentication
Authenticate receipt?	No: encryption contains no mechanism to prevent repudiation by a receiver	No: not in basic authentication, which authenticates only sender or receiver identities
Processing speed (delay)	Minimal processing delay; often performed in hardware	Greater delay due to processing of public key and private key
Ease of key management	Complicated without the assistance of public keys	Simpler when portions of keys can be distributed in public
Cost	Hardware is more expensive than software implementations; generally more expensive than authentication	Roughly equal to encryption costs in hardware implementations; much cheaper in software than encryption
Compliance with standards	DES (current); Skipjack (proposed)	DSS (federal); RSA (de facto standard)
Signature	Key value is a crude signature	RSA and DSS are intended to be electronic signatures
Hashing	Not a part of encryption	Secure Hashing Algorithm (Digital Signature Standard)
Compatibility with archiving	Works well with archiving systems due to speed	Often inadequately secure and too slow for archiving

Bulk voice encryption devices can encrypt T-1 or T-3 lines, with products offered by Cylink (Sunnyvale, Calif.) and many T-1 channel service unit vendors. Bulk encryption is very cost effective because it encrypts everything that is carried over a T span, including voice, data, facsimile, frame relay, ISDN, and Signaling System 7 traffic. While more of these lines are fiber and hence harder to tap, there remain many points at which T-span signals exist in electronic or microwave radio-frequency form. Since the T-span data formats are all standard, decoding them is very easy.

The Future of DES

DES's viability has been reviewed and renewed several times since it became a FIPS in 1977. DES was reviewed in 1988 and it was extended as a standard until 1992. DES has since been renewed yet again until 1998.

A recent study of DES and the new Skipjack algorithm by five experts indicated that at the current level of technology, a DES system could be broken in 10 years compared to Skipjack's 36. One reason is the *key space,* or size of the key. The more bits, the more possibilities. DES has 2^{56} or a little more than 7×10^{16} possibilities. Although extending the key space would give DES added life, the effect is not as pronounced as one would think.

The planned replacement for DES (although the government has not said so in so many words) is the Skipjack algorithm. Skipjack has an 80-bit key space, but uses an altogether different (and secret) algorithm.

"DES in its basic mode is very close to the end of its useful life," says Martin Hellman, a co-inventor of the competing public-key method. (Consider the source, too.)

Skipjack

Contrary to usual practice, the Skipjack algorithm has not been published. It is being kept secret by embedding it in silicon chipsets. While disclosure would not compromise it, the object is to deny the powerful algorithm to current or potential adversaries and keep its Law Enforcement Access Field secret.

In an effort to placate those who were outraged at being kept out of the loop, NIST commissioned five experts to examine Skipjack. The experts found Skipjack secure. However, this sop did not placate the encryption community by any means, and for good reason as we will see.

The Clipper chips

Clipper is the code name for a chipset using the new Skipjack algorithm (see Fig. 5.5). Clipper is intended to be used in the public switched data network for voice, data, and facsimile. Clipper is a private-key system, and the choice of the Clinton Administration for government transmission of unclassified but sensitive information. The administration also wants the private sector to use Clipper.

Unlike DES's evolution, there was no industry input into Clipper's development, which was performed by NSA. Although AT&T backs Clipper, other industry giants such as Motorola have criticized Clipper because it is not an open technology. Their reason is that all the chips (designated MYK-78's) are made by a single vendor, Mykotronx of Torrance, Calif. Working under strict NSA guidelines, they have so far built some 20,000 Clipper chipsets.

Clipper will not foster interoperability because signaling controls and voice encoding are not addressed by Clipper. While AT&T offered to license its technology (price not mentioned) to others for this purpose, Motorola claimed that generic (free) standards are available.

Speed is also a question. NIST's proposed encryption standard discussed Clipper in terms of the ISDN 2B + D speed of 2×64

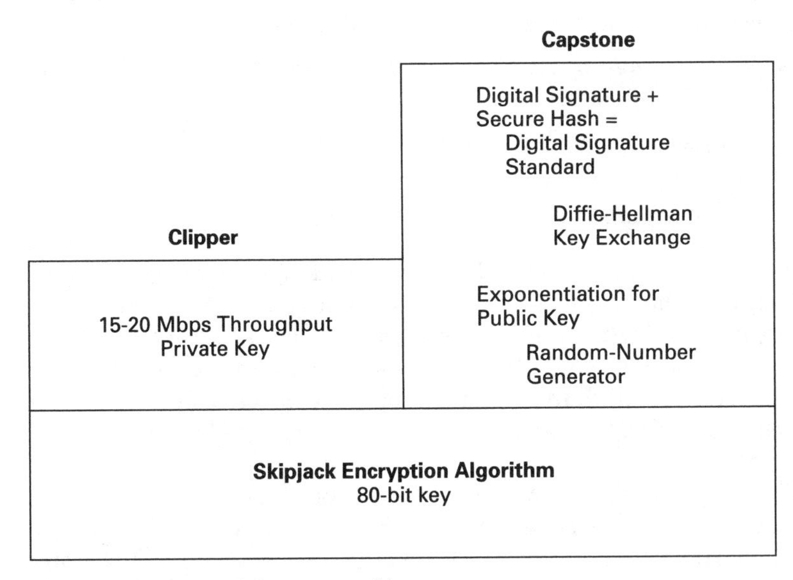

Figure 5.5 Clipper and Capstone architecture.

kbit/s + 16 kbit/s. Yet NSA said that Clipper chips will run as fast as 15 Mbit/s.

Application programming interfaces

NIST has issued a set of high-level application programming interfaces (APIs) intended to allow vendors to couple their applications software to any encryption algorithm: DES, Skipjack, or RSA. Furthermore, the APIs would work with PCMCIA cards, smart cards, or software. Being standard, any branch of government could use any combination of vendors' products.

Some vendors, including Apple, Lotus, and Novell/WordPerfect already use APIs based on the open Public Key Cryptography Standard. However, the government may want changes that more fully suit its needs. Once approved, the APIs will become a mandatory purchasing standard for U.S. government agencies.

Capstone

Capstone is a code name for an enhanced Clipper chipset, the MYK-80. Relying on the same Skipjack algorithm, Capstone adds the following:

- The Digital Signature Algorithm and the Secure Hash Algorithm, together called the Digital Signature Standard. These provide signatures and data integrity verification.
- A key exchange algorithm based on a technique developed by Whitfield Diffie and Martin Hellman.
- An "exponentiation" algorithm to provide the public-key function.
- A random number generator.

You can see that Capstone adds source and data integrity authentication, plus the flexibility of public-key distribution.

The encryption community's apparent concern over a lack of public review became real when Matthew Blaze, a researcher at AT&T Bell Laboratories, found a way to prevent government decryption of Tessera (a PCMCIA card containing a Capstone chip) encrypted output.[4]

The Skipjack algorithm embedded in the Capstone chip contains a Law Enforcement Access Field (LEAF). This LEAF field allows the government to decrypt a Capstone-encrypted message using the escrowed keys discussed earlier. Blaze built a LEAF field by binding a false session key into the LEAF that is accepted by a recipient Capstone chip but does not contain the encrypted session key that the government needs to decrypt the message.

Blaze's accomplishment contains three ironies. First, it revalidates the need and desirability of public review. Public review encourages examination from all viewpoints and compensates for competing entities' (i.e., government or industry) vested interests, since no two entities will have the same axe to grind.

Second, AT&T sold the FBI some 8,000 Secure Telephone Devices based on the Clipper chip that uses the same Skipjack algorithm their own researcher broke. (AT&T later claimed that a *New York Times* article erred when it said that Clipper could be compromised in the same way Blaze beat the Capstone chip.)

Third, NSA refused to release the Skipjack algorithm for public review for fear that the LEAF filed would be jeopardized. It is now clear that disclosure was not necessary in order to imperil the LEAF field.

Trusted Information Systems of Glendale, Md. has proposed two enhancements to the LEAF. To thwart the Blaze attack, both ends of the conversation compute the LEAF using the chosen session key. After so authenticating, the receiver would lock up if the compared LEAF later failed to match. Intruders would be required at both ends to foil this approach, so such a breach is considered unlikely. The other enhancement is to allow both the LEAF as well as public or private keying. The LEAF would be preserved even if private or public keys are used.

Military origin

Both Clipper and Capstone were designed for the Department of Defense's planned Defense Message System. The Defense Message System type 2 algorithm is in fact the Clipper algorithm.

As mentioned above, Capstone chips will be mounted in credit-card sized PCMCIA cards called *Tessera* cards. Tessera cryptocards will be used to enter and turn on Capstone's public key and signing attributes in the Defense Messaging System.

A key escrow system

Vendors selling Clipper- or Capstone-equipped devices must send a copy of the key used in each machine sold to the Department of Justice and NIST, who will act as the escrow agents. Keys for Clipper and Capstone are to be held by these two federal agencies, half by each. A court order would be needed to obtain the key halves needed to tap a line. The reasoning is that dishonest persons can use Clipper or Capstone to hide their evil machinations. The government needs a way to tap them.

One problem with the escrow system is that both agencies present a target for theft, regardless of whether the key halves are kept in a database or a black box of some kind. The keys have no time limit. The government could decode any message ever made by the Clipper or Capstone chipset in question, before, during, or after a warrant expires.

There is also a Fifth Amendment implication. If you have told no one the combination to your safe, then the government cannot compel you to disclose it. If you have disclosed it to a third party, then the government can require disclosure. Does escrow constitute a third party and a loss of Fifth Amendment rights?

As Kate Martin, an attorney with the American Civil Liberties Union, said, "[The] Clipper chip should not be called a key escrow system. It should be called a system in which the government holds the key. You won't be able to sue the government for a breach of fiduciary responsibility without a change in the law." The Justice Department has voiced the same position: Individuals have no right of recourse against the government if it fails to protect the keys it holds.

Justification?

The government justifies the program on the basis that they need to be able to tap suspected wrongdoers. In fact, F.B.I. Director Louis Freeh has called encryption an obstacle to law enforcement several times in testimony before Congress. What is to stop a wrongdoer from simply using DES? The implication is that the government may have a long-term plan to bar the use of DES or any other form of encryption but Clipper or Capstone in the United States. Some government officials have stated this objective directly, but others have dismissed it as not being an option. The White House Office of Science, Technology, and Policy has defended Clipper's policy as having been reviewed at the highest levels.

One government has already taken action. In April 1995, Russian president Boris Yeltsin decreed that the use of encryption hardware and software is illegal without a license. Details of the edict can be found on the World Wide Web at http://www.nww.com/ruscrypto.html.

An option that was suggested by Vice President Gore in July, 1994 is the development of an unclassified cryptosystem for private industry. While such a system would not replace the Clipper chip key escrow system, it would be subject to public review. It would, however, also be an escrow arrangement, a move opposed by public interest groups such as the Electronic Privacy Information Center. The move is clearly an attempt to appease the computer and communications industry, which has almost unanimously opposed the Clipper chip system.

Government has never before intruded in private use of encryption. That it is doing so now is of great concern, especially since its motives are unclear. Its motives became even less clear when it said that more algorithms of an unspecified nature might follow.

Thomas Jefferson said that mistrust of government was essential to a free society. Two centuries later, John Perry Barlow, a co-founder of the Electronic Frontier Foundation, said, "Cryptography...is the 'material' from which the walls, boundaries—and bottles—of cyberspace will be fashioned." If cyberspace is one dimension of a free society, then the ability to secret information from prying government eyes is a requisite.

Clipper does have some legitimate uses. It will be usable in the forthcoming National Information Infrastructure to protect data coursing down the nation's information highway. Electronic commerce, being pressed as a way to help American business and government streamline their operations, cannot function unless information sent electronically can be validated. In fact, NIST has proposed an Electronic Commerce Integration Facility (ECIF) for just this purpose. ECIF would be composed of government agencies and perhaps vendors as well. In the public sector, the health care reform package relies heavily on standard electronic forms processing and routing.

There are some nagging indirect issues:

- Is government control of these chips appropriate?
- What will the availability, quality, and price be?
- There is only one source now. Despite AT&T's support for Clipper, AT&T has declined to be a second source.
- What will foreign reaction be? Will retaliation occur? American companies are all over the world. Foreign reaction thus far has been overwhelmingly negative.
- Will exportability be affected? Some foreign companies are our manufacturing partners. Can chips be installed legally overseas?

- Clipper or Capstone meets no international or banking standards. Since our overseas allies were not consulted, their cooperation will be minimal. Resistance has already surfaced from the banking community, which presently uses RSA-based security and is loath to change.

Staying informed

If you are interested in the Clipper and Capstone issue, you can bring up the Risks forum or the Privacy forum on the Internet. There is an alt.privacy.clipper news group. RSA Data Security, Inc. is accessible on-line at rsa.com.

Review

- *Cryptology* is the study of ways to disguise information so that only the intended recipient can understand it.
- *Plaintext* is converted into *ciphertext* in a process called *encryption.*
- An encryption *key* allows the cipher stream to be changed at will.
- Changing the key is one protection against adversarial *cryptanalysis.*
- Both symmetric *single-key* and asymmetric *two-key* cryptosystems exist.
- Single-key cryptosystems include *authentication* automatically.
- Two-key systems such as RSA or DSS may perform encryption and/or authentication.
- The *Data Encryption Standard* or *DES* is a single-key system.
- DES became *FIPS Standard 46* in 1977. The algorithm is *public.*
- DES uses a *pseudorandom stream* of bits mixed with text to form a *ciphertext.* The DES key is *56 bits* long.
- DES may be used in *WANs* and *LANs.*
- *Key management* is the major problem in big networks; ANSI X9.17 sets standards for key management. Key management centers are used in large networks.
- *DES may be implemented in hardware or software,* but hardware is faster.

- *The RSA two-key system is more convenient and flexible,* but slower and more expensive.
- *DES's lifetime is slowly coming to a close:* perhaps in the late 1990s.
- It may be replaced by a new algorithm, *Skipjack.*
- There should be a major public debate before a *key escrow system* such as *Clipper* and/or *Capstone* is accepted, if it ever is.

Appendix. Vendors

Here is brief listing of vendors offering the technology discussed in this chapter and in the notes.

AT&T Secure Communications
Customer Service Center
(800) 203-5563
Offers SecretAgent Software to provide DES, DSA, RSA, and Secure Hashing Standard applications. DOS, Windows, Macintosh, and assorted UNIX platforms are supported.
Surity DSA Signature Software offers only Digital Signature Algorithm (DSA) and Secure Hashing Algorithm (SHA). DOS, UNIX, and Windows platforms are supported, with Macintosh in development.
Cryptographic Development Kits are a threefold library of linkable code modules for software developers who wish to add security into DOS, Windows, Macintosh, and UNIX applications. One kit includes DES, ElGamal public key, DSA, and SHS. A second includes most RSA functions. The third includes RSA and DSA kits.

Enterprise Solutions, Inc.
(818) 597-9621
Offers encryption and authentication enhancements to X.400 enterprise mail systems.

Fischer International Systems Corp.
Naples, Fla.
(800) 237-4510
Offers WorkFlow.2020 forms application software. Also sells ARMOR ISA-PC circuit boards using Capstone to create DSS signatures and Skipjack encryption through AT&T. Capstone is also offered in software form as SmartDisk for signing and encrypting electronic documents.

General Magic, Inc.
(415) 966-6721
Licenses RSA technology for its TeleScript communications product.

Hewlett-Packard Co.
(408) 447-4600
Offers a Cryptographic Security Module for HP 9000 Series and 800 Series Business Servers.

Information Security Corp.
Deerfield, Ill.
Develops SecretAgent for AT&T.

National Semiconductor Corp.
(408) 721-2253
Sells iPower integrated-circuit-based encryption products

RSA Data Security Corp.
Redwood City, Calif.
(415) 595-8782
Toolkit for Interoperable Privacy-Enhanced Messaging: a software toolkit offered to develop applications using digital signatures and encryption.

Semaphore Communications Corp.
(408) 980-7751
Offers a Network Encryption Unit that supports DecNet Phase IV, VAX Cluster, PathWorks, and AppleTalk protocols among others.

Notes

1 *Security Requirements for Cryptographic Modules,* FIPS Publication No. 140-1, available by calling the National Technical Information Service at (800) 553-6847 or (703) 487-4650.

2 *The Public Key Infrastructure Study,* MITRE Corp., Vienna, VA, 1994.

3 Michael Baum, *Federal Certification Authority Liability and Policy,* Independent Monitoring, Cambridge, MA, 1994.

4 A copy of a paper written by Matthew Blaze that describes his methodology is available on the Internet. Use anonymous ftp and connect to research.att.com. Using cd, switch to the dist/mab directory. The Postscript-format file is titled eesproto.ps.

5 Bruce Schnier, *Applied Cryptography,* Wiley, New York, 1993, and Gilbert Held, *Top Secret Data Encryption Techniques,* Sams Publications.

Chapter

6

Access Control

Objectives

At the end of this chapter you will be able to:

- Know how to build strong passwords
- Understand the perimeter strategy and how to implement it
- Comprehend the underlying principles behind authentication
- Distinguish between the security roles of user IDs and passwords
- List and explain the four basic principles behind password management
- Understand the role of auditing in controlling access
- Explain the proper role of the system administrator in access control
- Evaluate hardware and software access control products
- Understand the security complexities presented by distributed networks
- Discuss security solutions for distributed and client–server networks
- Outline the functions of Kerberos software, along with its pluses and minuses
- Recognize and seek out security holes in your organization

Access Control

The most effective security is often the simplest. Almost without exception, the products and services we buy today have some security features, often far more than we really need. Ironically, the rich blend of security mechanisms can work against us. Instead of implementing them, we are intimidated by all the nuances, bells, and whistles. In our haste to get the application running, the complexities of security are given short shrift.

The first part of this chapter concentrates on one of the most powerful access control devices we have, passwords. They cost so little, yet are abused in so many ways that their security value is greatly diminished. So our initial focus is on correcting these abuses.

In the second portion of the chapter, we look at software and hardware access control mechanisms and their strengths, weaknesses, and applications. Perhaps the most challenging application from a security standpoint is the very dynamic client–server computing environment, for which special security measures are required. We will illustrate the Kerberos software solution to this problem at the chapter's conclusion.

Security used to be much easier when we had centralized mainframes and stiffly hierarchical networks. However, today we are building distributed networks as fast as we can. Many of these single-layer networks are tied together by long-haul telephone lines or satellites. Furthermore, our user community is very mobile and needs access to ever-more online resources.

The major contradiction is that we are building more open systems permitting greater access...to a point. Knowing where that point lies is the rub. Accordingly, the security needs of client–server networks vary somewhat from those of a conventional LAN. Further, new technology exemplified by the distributed computing and management environments calls for a new security approach. Called *Kerberos,* we will look at one way to manage security in an open client–server environment.

In an effort to give you a head start on discovering and filling security holes, we will give you some examples of what to look for. Consider your own environment as you read. You may wish to take some notes.

Spending money is always a popular way of enhancing security. Filling a security hole with cash, though, is not a wise strategy.

We will look at some access control products with the specific intent of giving you guidelines for evaluation, rather than discussing one product versus another.

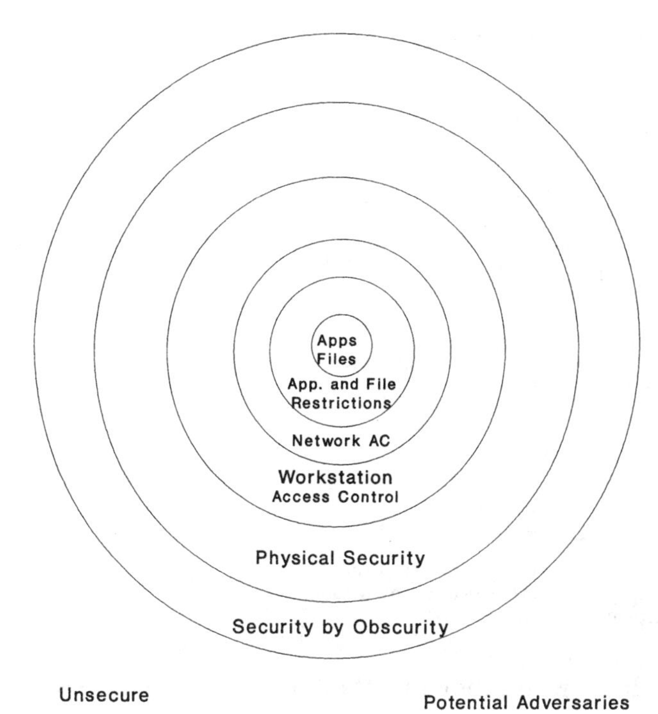

Figure 6.1 The security perimeter.

The Perimeter Approach

As Fig. 6.1 shows, access control begins with physical security. If you can keep would-be intruders out, then the inner defenses are not necessary. In the words of one planner at a midsize, multisite company, "Take a perimeter approach. Start with a very large perimeter and secure everything around that perimeter. Then you gradually start shrinking your way in, making it [the perimeter] smaller and smaller." By the time we get to network authenticators, we are already working inside a fairly small perimeter.

Individual Authenticators

In Chapter 4 we discussed the concept of authentication. A pictorial view is shown in Fig. 6.2.

User IDs and passwords are both type 1 authenticators: authentication by knowledge. If we ask a user to enter a password after card-reading their badge, we have added type 2 ownership to the

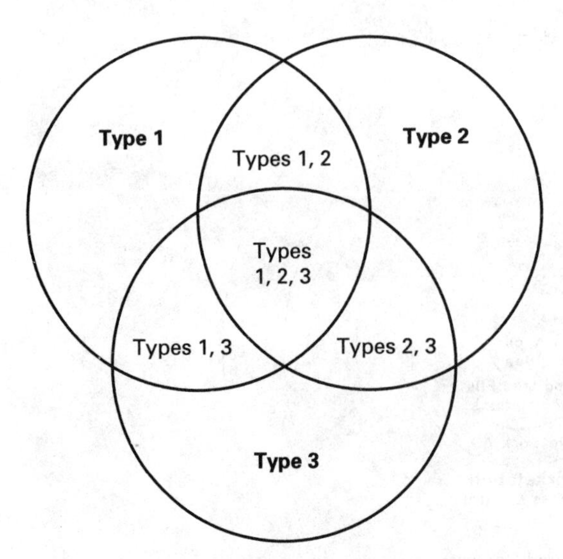

Figure 6.2 Authentication methods. Type 1: authentication by knowledge; type 2: authentication by ownership; type 3: authentication by characteristic.

formula. If we do so while they are standing on a scale, we are using all three methods.

The principal weakness of types 1 and 2 is that the user is not physically attached to either. So the password and badge could be lost, stolen, or otherwise compromised. Worst of all, passwords can be duplicated. A crook may look over your shoulder. Crooks do not even have to do that: they can guess. The guessing game can range in sophistication from no special tools, skill, or equipment to computerized brute force attacks employing demon dialers and Cray machines attempting to crack Data Encryption Standards (DES) keys. Two advantages of a password are that the user can take it anywhere and that the password is easily changed, whether periodically or in case of compromise.

Something owned (type 2 authenticator) such as a badge or an ATM card is hard to duplicate. Passwords are "owned" in a sense and so are a subset of type 2. Authentication by characteristic (type 3 authenticator) is a much stronger form of authentication than either types 1 or 2, because the characteristics that make you unique are inseparable from your identity.

Fortunately, it is not necessary to measure all the characteristics of a person to establish their identity. A fingerprint or a handprint may be enough. Just *one* aspect of characteristic measurement is often sufficient. This is fortunate because any kind of biometric measurement is more expensive than a password.

We can see then that placing the type of authentication in descending order of security value is type 3, type 2, and type 1. Combining types 1 and 2, as is often done, approaches type 3 assurance.

User IDs

A user ID is a unique symbol or character string used by a system to uniquely identify the user—a claim of identification to be authenticated by a distinct password. User IDs are not intended to be secret.

A user ID should be assigned to only one person. Group user IDs for nonauthentication purposes (e.g., mail) are permissible. However, if two or more people know the password for a given user ID, that *is* a security violation.

In a relative sense, a user ID is an outer perimeter security tool relative to a password. They share many of the same characteristics that make them security fences to be hurdled.

User IDs should be *blacklisted* after a certain number of tries. Usually the number will be large, about 12 attempts. Then some kind of contact between the system administrator and user should be necessary to reestablish privileges. While adversaries can try another user ID, this takes additional time. The objective is to make even a protracted attack over time yield as few user IDs as possible. Locking out a user ID whose associated password fails to authenticate after a dozen or so attempts is an outer perimeter safeguard. To this end, it is advisable to use as many different user IDs as possible.

Passwords

> If everyone were to use a really tough password and keep it secret, 99 percent of our problems would go away. DAVE JOHNSON, *Consultant, Talon, Los Altos, Calif.*

Principles

These are the core principles for dealing with passwords:

- Users should be able to change their own passwords.
- Passwords should be machine, not user-generated.
- An audit usage report should be provided directly to the user at least monthly. As a short-period check, each new login should

include the date and time of the last login under that user ID. Some newer systems give an on-screen audit report of logins within the past week or even monthly, eliminating the need for a separate monthly report while providing at the same time a rolling 30-day audit report.

- Automation increases security. Use it wherever possible.
- Passwords are not absolute providers of security. They may only be strong enough to stop a casual intruder, not a determined hacker.

Passwords *authenticate* the identity of a user. They verify the claim made by a user ID. Password administration is summarized in Table 6.1.

Initial system passwords

New systems often contain default passwords that are well known by now: e.g., System, Master, Test, Guest, Field Service, and the like. Change or remove all of them upon installation. The trend in the last year, especially for PBX vendors, is to ship systems with *no* installed passwords. Instead, the user is asked to enter passwords of their own choosing as needed to initialize the system.

Initial password assignment

The system administrator should not know a user's password. If this is not possible, then the user should change the password as soon as possible after the initial assignment. It may be necessary to force the system to perceive the password given the user as expired in order to force an out-of-cycle change. Another method is for the system to allow one-time use of the given password, then insist on a new one before the next session.

Password length

The Department of Defense recommends at least six characters. Eight characters are common on many systems. All characters should be usable. Four factors affect password length:

- The maximum lifetime (L) that a password can be used
- The probability (P) that a password can be guessed during its lifetime, assuming that the guesser works continuously
- The number of guesses (R) per unit of time that it is possible to make

TABLE 6.1 Password Summary

Password generators	Administrator User An automated computer routine Some combination of the above
Distribution	Mail Verbal Hand-carried CRT display
Recording methods in the computer	Clear text Encrypted Algorithmically recorded
Recording by user	Written (to be discouraged) Memorized Written via an algorithm
Construction	Use the full character set, alpha, numeric, upper- and lower-case characters for maximum protection against an organized attack. Greater complexity increases the likelihood that the password will be written down. This may mitigate in favor of simpler alphanumeric passwords if you accept the notion that an organized attack is less of a risk than a compromise due to one or more passwords being written down. Password construction logic options include random, dictionary words, phonetic, and phrases.
Changes	User Administrator Automatic routine ("Your password is about to expire. Please enter a new one now.") In response to termination, transfer, resignation, or compromise
Exposure	Administrators Programmers (e.g., via system dumps) Password record observation Shoulder surfing

- The password "space" (S): the total number of possible unique passwords

The four are related by a simple formula:

$$P = \frac{L \times R}{S}$$

Further information on numerical computation of password lengths is found at the end of the chapter.[1]

Passphrases

A passphrase is a concatenation of words drawn from a dictionary, such as manbat. Using four-, five-, and six-letter words, about 23,000 passphrases can be built. Words can appear more than once in a passphrase.

Passphrases are easy to create and remember. The more concatenation, the more security. They also provide a degree of randomness beyond what an individual would likely select.

Phonetics

A password may also be phonetic: pronounceable but meaningless. Examples include Jisked or Widlow.

Password changes

If a user forgets his or her password or decides that it has been compromised, the administrator should be able to assign an arbitrary new one without knowing the old one. Then the user should change the new password to make it private.

Revalidating IDs

A mechanism must be in place to remove a user ID and password when the employee resigns or is terminated. User IDs (and therefore, indirectly, passwords) should be revalidated periodically, at least twice a year and preferably every 60 to 90 days. The name of their sponsor and the user's phone number and address should be updated if necessary.

User responsibilities

Users must understand that they are partners in maintaining security. They must keep passwords secret. They should be told to report changes in job or access status or possible violations they see or suspect. Some companies have employees sign a statement acknowledging their security awareness responsibilities. Consistent with this commitment, a security awareness program is a good idea and its formulation is discussed in Chapter 12.

Changing passwords

Passwords should be changed periodically, at least twice a year. Some systems force users to change their passwords more often, typically every 60 to 90 days. They become stale because sometimes other people see or learn them. The more time, the more suspect the password must inevitably become. Remember that password changes do not protect you from an adversary who guesses at passwords. This probability is a constant, regardless of how often a password is changed.

Since users should change their own passwords, they should login with their old user ID and password. Once authenticated, they should then change their password to something new. Users should be allowed to change only their own passwords and never anyone else's.

Passwords should be *remembered, not written down*. Saying this does not work. Instead, let users select a password from an automatically generated phonetic or passphrase list. Any password on the list is by definition pronounceable and a more random choice than a user would likely select.

Many, many studies show that penetration is harder when an automated password selection process is used. While it is tempting to use all available characters, users resent the resulting complicated IDs and passwords. There are a number of password generator programs available to create pronounceable user IDs and passwords.[2]

Login attempt rate

If you can control the rate that password attempts are made by a given user's port, then do so. That is, you should delay another attempt after a failed one. The rate should be between once a second and once a minute.

The system should report five or so unsuccessful login attempts from a given port or user ID. In and of itself, such a warning is not too meaningful: maybe the user is a bad typist. However, repeated warnings may imply that someone is trying to break in. Accordingly, some systems will lock a user out after a number of attempts. A common limit is three.

If an adversary can switch from one terminal to another quickly, then the maximum user ID attempt rate should be reduced as well. Put another way, the mandatory waiting period between attempts should be increased.

Audits

An audit trail should contain a listing of password usage and changes. It should not contain actual passwords.

Events that should be audited include

- Successful logins
- Unsuccessful login attempts
- Use of the password change procedure
- User ID lockouts due to attempted use of an expired password
- Date and time of events
- The offered user ID for unsuccessful logins
- Actual user ID for other events
- Origin (e.g, location or port ID) of the event
- Successful and unsuccessful password changes by a given user ID

User audit

Upon logging in, all users should be given:

- The date and time of their last login
- Their location (as best can be determined)
- Any unsuccessful login attempts to the user ID since the last successful one

Users should be told at each login to examine this brief audit statement and report any suspected discrepancies (i.e., a login attempt while the user was away on a business trip or on vacation).

Access Control Hardware

Beyond remembering a password, there are systems that call for both physical possession of a device (a "physical token") and a password that may vary by time or even be confirmed biometrically. Of the three looked at here, two are strictly authenticators while the third includes encryption.

Dynamic password control

One form of smart authenticator is a credit-card sized device[3] carried by the user. By pressing a button, the liquid crystal display on the card shows a sequence of eight numbers. The numbers change every 60 s and are the valid password for the period. Optionally, the card will ask the user for a personal identification number (PIN) before doing anything.

Security software in the host computes the same numbers on the same schedule. A remote caller must enter the right code for that brief period or else the login attempt will fail. There can be several valid codes at once, unique by user. This allows the system to automatically track usage by individual.

Beyond manual use, dynamic password client software can be implemented in another host, in a PC, or in any other kind of workstation. Host (central) systems can be a server or a separate access control module. The access software can also be installed in a mainframe or minicomputer. This way, even if a tapper records a password, it becomes meaningless 60 s later. Such devices are strictly authenticators: they offer no protection for data.

One risk in such systems is that a "tailgater" can piggyback themselves onto an authorized call. In the words of Lenny DiCicco, convicted of breaking into a DEC network, "...once the user had validated the login, we'd seize control of the line, dump them off, jump onto the line, and take over where they left off."

Dynamic password systems are implemented as remote dial-in (using an AT command set string) for PC and Macintosh users and are also implemented for pushbutton (Touch-Tone) control systems. The author saw one such application regulating access to a cellular telephone switch made by Ericsson. Clearly, access to the switch by a hacker could be disastrous, but legitimate card-carrying maintenance technicians could access the switch's maintenance port from any telephone location.

Workstation authenticators

Another system[4] uses an Industry Standard Architecture/Extended ISA (ISA/EISA) controller card installed within a PC workstation. A small card is programmed with data identifying the user and approved workstation(s), then inserted in a slot that looks like a disk slot. The controller checks for validity before allowing workstation access. The workstation is disabled when the card is removed.

A variation by another vendor[5] adds host computer port protection and DES encryption for PCs and Unix workstations. The card can double as a badge or other security identification. Even biometric data can be loaded on the card.

Access Control Software

Access control software controls access to the workstation before a user can get to any workstation-based applications, much less get to a network. These products establish yet another perimeter around the network. Access control functions include identifying authorized users and limiting access to files, applications, devices, and LAN segments.

These tools use logins, passwords, and sometimes firmware boards or smartcards and card readers. They are intended for workstation use only, and are not server-based products.[6] Basic evaluation criteria are shown in Table 6.2.

Security in Distributed Networks

In multiserver or multihost networks, the security challenge is twofold. Often the user logs in from place to place, so we must authenticate the person, not a place. Second, the user requires access to some resources but not to others.

There are two basic solutions. One is to have all users log into a security server first. From there, the server controls usage of resources. This is the principle behind the Kerberos scheme used in the distributed computing environment (DCE) of the Open Software Foundation (OSF).

The second solution is more difficult. The user logs into a particular machine and receives services as specified. To go elsewhere, the user repeats the process. This is a typical environment for many Novell NetWare 3.X users. In such a system, the administrator must manage the number of servers multiplied by the number of users. The administrative workload can quickly become untenable.

Those in such a position are advised to consider upgrading to NetWare 4.X or installing Banyan Systems' Enterprise Services for NetWare. Either approach consolidates security into one location

TABLE 6.2 Access Control Software Evaluation Criteria

Ease of Use (E.g., Menu vs. Command-Line Interface)	
System criteria	Does it modify system files? (It should not need to) Deny access until boot and access program load is complete Prevent booting from a floppy disk, thereby bypassing security Deny rename, deleting, or overwriting of environmental files such as config.sys, command.com, hidden files, or a security program's executable files
Levels of access control allowed	User ID and password for the network manager, plus multiple levels for users The product must support multiple users (at least ten)
Passwords	Settable minimum password length Allows at least 10-character passwords Supports all 256 extended-ASCII characters Allows password expiration times Has password strengtheners: expects upper- and lowercase; special characters Warns of weak password selection
Audit functions	Controls illegal access attempts: locks user out of workstation Preserves an audit trail in battery memory or protected file Displays attempt information on monitor Audit trails: many have too much detail Time stamps beginning and end of all sessions Has user authentication index, ID codes, and utility action code used by manager or user (proves that user in audit trail is the correct user) Has a program and file usage trail Has an encrypted audit log: may be hidden file or read only by end user Informs user of time and date of most recent login
Antisurfing	Inactivity timer to protect data on screen or in memory

(as far as the user is concerned) and vastly simplifies the security task.

Distributed networks imply wide area network (WAN) connections, often with bridges and routers. Security issues related to these items are discussed in Chapter 9.

Client–Server: Kerberos

Kerberos is authentication software that can be added to an existing security system. It authenticates the user as legitimate to the system and the specific application or service sought. Since all its exchanges are encrypted, intercepting and analyzing network traffic are useless.

Applications do not generally run easily with Kerberos. It can take a programmer a day per application or more to make them work together.

The Open Computing Software Group of Redmond, Wash. specializes in Kerberos software. They offer Unix, Macintosh, and PC versions. IBM and the Open Software Foundation in Cambridge, Mass. plan to make Kerberos part of their Unix operating systems.

How Kerberos works

If you are not up on your Greek mythology today, *Kerberos* was the three-headed dog who guarded the gates to Hades. The analogy is that the Kerberos system uses a database, an authentication server, and a ticket-granting server to guard the network (see Fig. 6.3). All of these run on a single physical server, hopefully locked up somewhere.

When a user logs onto a workstation, he or she requests service on the host. The authentication server checks the database seeking the user's access rights. Finding them, the authentication server creates an encrypted ticket to let the workstation access the ticket-granting server. It also returns a copy of a randomly generated session key used to encrypt the *authenticator.*

The authenticator grants users access to specific services. It makes sure that the person is really the person who was granted that ticket. The workstation then sends the encrypted ticket and the authenticator containing the user's name, network address, and present time to the ticket-granting server.

The ticket-granting server decrypts ticket and authenticator. If they agree, the ticket-granting server creates a ticket for the host (server) that the user seeks to access. Upon presentation of the

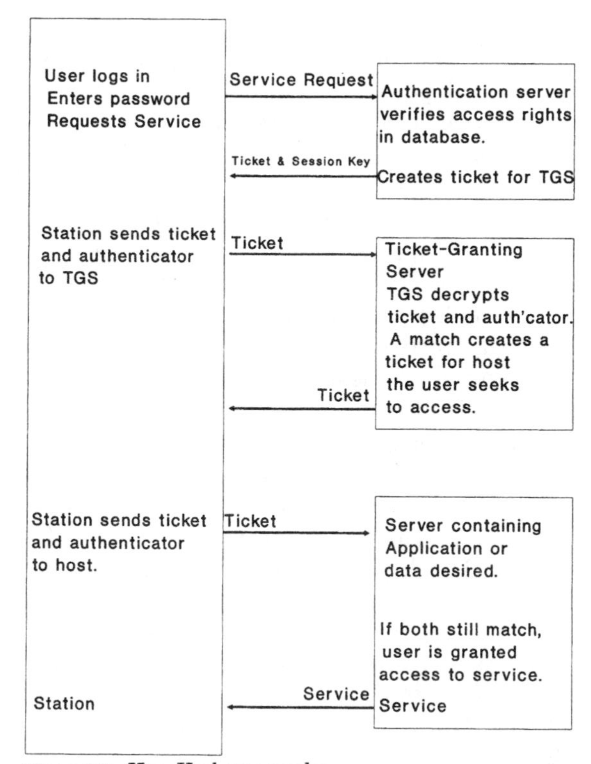

Figure 6.3 How Kerberos works.

ticket to the server, the application can be run or data can be accessed.

Discussion

Though a strong system, Kerberos has its weak spots. One has already been mentioned: every network application must be modified to work with Kerberos. Thus one must have the application source code and invest significant programming resources. Leased off-the-shelf software is almost impossible to adapt to Kerberos.

Since Kerberos relies heavily on passwords as the only way to identify a user, Kerberos is subject to a dictionary attack where an attacker takes all the words in a dictionary and DES-encrypts them. The attacker then compares the encrypted words to a stolen network password, that is, of course, encrypted itself. A match yields the plaintext password.

Avoiding dictionary words invalidates this kind of attack. As noted above, a passphrase or phonetic password is good. Mixing numbers and letters is also good: perhaps, "Not2Bad."

Maintaining Kerberos service in the servers and the overhead spent on encryption are, respectively, administrative and performance issues. Network overhead is a potential problem, the point being that several tickets should be issued for a session and then remain valid for hours before the process needs to be repeated.

On the plus side, Kerberos is well suited to distributed networks and client–server networks. It clearly offers more protection than a simple password. Best of all, Kerberos is transparent to the end user. After the user enters an initial password, it is automatic, taking most of the security responsibility out of the user's hands.

Kerberos is available under license from MIT, where it was created in 1986. In 1989, MIT offered Kerberos to the Open Software Foundation for its distributed computing environment; it was accepted. Accordingly, Kerberos is now incorporated in DCE software. Vendors such as IBM, Hewlett-Packard, DEC, and Sun all plan to support it.

IBM offered the OSF a variation of Kerberos called KryptoKnight, which uses authentication and distribution protocols claimed to be more resistant to attack. While Kerberos uses a client–server architecture, KryptoKnight allows initiators and responders be peers, making the system easier to manage.

IBM is also working on Security and Management Service in Open Networks (SAMSON). SAMSON provides security services and mechanisms for any kind of network. It uses security as specified in OSI. SAMSON can be used to define secure objects and is intended to work with the DCE and ITU-TSS X.500 Directory Services.

Another vendor, Northern Telecom, Inc., offers a public-key security system called Entrust. Entrust has both digital signature and encryption capability. Uniquely, it solves the key distribution problem by using the ITU-TSS X.500 Directory Service. Public keys are kept on a server to which anyone has access. The X.500 Directory Service allows multiple servers to be spread around the enterprise and so scales nicely. Furthermore, one X.500 service has the management tools needed to allow administrators to manage different kinds of data such as login files and, of course, public keys in a standard way.

Application developers use an Entrust API to place security applications atop other applications such as workflow software and groupware. The API can also be used to encrypt individual documents. Entrust has found its way into some major client–server applications, namely Lotus Notes and Microsoft's Exchange client–server messaging product.

Summary

Where to look for security holes

Holes relating to access control include:

- Failure to establish perimeters
- Weak passwords
- Old passwords retained from original system installation
- Passwords not removed
- Writing passwords down
- User-selected passwords
- Sharing passwords
- Insensitivity to security
- Too much trust in dial-back

Review

- *Access control* is one of our most powerful security techniques, but is often abused.
- User IDs and passwords form the core of the *perimeter* approach.
- There are three basic types of authenticators: *knowledge, ownership,* and *characteristic.*
- Authentication methods can be meshed for greater security.
- User IDs are not a security authenticator.
- Passwords should be changed periodically and only by the user.
- Passwords should be machine generated.
- Audit reports play a key role in security management.
- Automation increases security.
- Users share the responsibility for security.
- Access control hardware is mainly for authentication. Fixed dial-back systems are particularly vulnerable to interception, despite smart card authenticators.

- Hardware authenticators represent another security perimeter.
- Implemented properly, access control software is equally effective.
- Distributed networks can perform security by server or in a dedicated server.
- Kerberos software uses a database, an authentication server, and a ticket-granting server in one physical box to provide network security.
- Application software will require modification to work, perhaps at great expense.
- Kerberos will be used in the client–server distributed computing environment.
- Plugging security holes is often a matter of common sense and knowing where to look.

Notes

1 Department of Defense Password Management Guideline, CSC-STD-002-85, April 12, 1985. Available free from NSA CMSC. Also see FIPS PUB 112, *Password Usage,* dated May 30, 1985. In part, it defines ten factors to be considered in building passwords. The document is available from NIST.

2 Morrie Gasser, *A Random Word Generator for Pronounceable Passwords,* ESD-TR-75-97, MTR-3006, AD-017676, MITRE Corp., Bedford, MA, November, 1975.

3 The vendor referenced is Security Dynamics, Inc.'s (Cambridge, MA) SecureID.

4 The vendor referenced is DataMedia Corp.'s (Nashua, NH) SECUREcard system.

5 The vendor referenced is SOTA-MILTOPE's PassPort CSS/OmegaCard.

6 Two well-regarded access control software products are Fifth Generation Systems' TriSpan and Pyramid Development Corp.'s PC/DACS.

Chapter

7

LAN Security

Objectives

At the end of this chapter you will be able to:

- List the do's and do nots of providing LAN physical security
- Show where LAN security fits in the ISO model
- Discuss the security implications of LAN wiring
- Explain how LAN analyzers can compromise security
- List electrical hazards to network security and how to combat them
- Explain how LAN hub address verification can augment security
- Describe the implementations of address verification
- Illustrate how verification can be used in Ethernet, token ring, and Fiber Distributed Data Interface (FDDI) networks
- List the advantages of hub-based security
- Show how terminal servers and configuration controls augment security
- Discuss encryption, authentication, and IEEE 802.10b protocol options
- Describe the use of bridges as security devices
- Outline TEMPEST considerations in the LAN context
- Identify security risks in using Simple Network Management Protocol (SNMP)

- Compare and contrast SNMPv1 and SNMPv2
- Discuss the security aspects of LAN inventory management
- Know what to look for in an inventory management software package

LAN Security

Local area networks are among the most vulnerable communications networks ever conceived. The inherent risk is the direct result of the unparalleled flexibility and connectivity that make LANs so useful. This chapter targets the opportunities for security breaches in the physical side of the LAN and indicates measures you can take to block possible leaks.

The network management dimension of security is generally all but ignored. Again, the very tools that make LANs manageable offer hackers another mark at which to aim. Mechanisms to bar this point of entry are discussed in this chapter.

One of the major barriers to overcoming network security risks in LANs is the fact that LANs are inherently decentralized. Security is thought of as a centralized function, directly contradicting the spread-out nature of local area networks.

Yet some aspects of LANs are centralized. Hubs are centralized, and inventory management is centralized. Those who claim that adequate security is limiting the implementation of LAN technology are mistaken. There is enough security hardware and software available to satisfy the most discriminating customer and drain the contents of the largest wallet.

This chapter concentrates on physical LAN security, including cabling and wiring hubs at layer 1 of the ISO Reference Model (see Fig. 7.1). Layer 2 security is implemented in LANs via frame [sometimes called hardware or network interface card (NIC)] address verification, and also via the IEEE 802.10b Interoperable LAN/MAN Security (SILS) Standard. At layer 6 we find encryption, data hashing, and authentication.

This chapter also deals with security aspects of the most widely used LAN management protocol, the Simple Network Management Protocol (SNMP). While SNMP has been widely accepted, it has not been fully implemented due in part to a lack of security controls. We discuss this failing and what has been done to fix it. SNMP version 2 has strong, flexible security and we will see how it works.

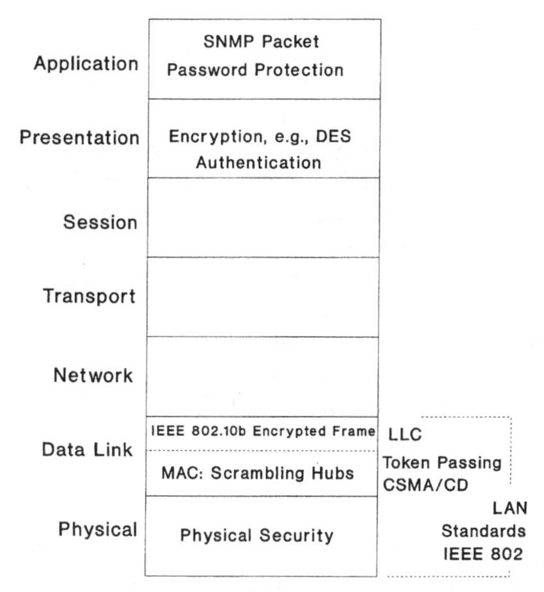

Figure 7.1 LAN security and the ISO model.

Inventory management is one element of network security, because we sometimes have the concurrent responsibility of tracking and accounting for all the hardware and software in the network. We are also vitally interested in workstation hardware and software configurations that might affect network operation or performance.

Network integrity can be undermined by computers that are stolen, added, moved, or modified. Inventory software that scans all machines on the network and determines what hardware and/or software is *really* in them is always an illuminating experience for the naive network manager. We conclude the chapter with a discussion of inventory software, what it can do, and how to evaluate the many packages on the market.

Physical Security

Once again taking the perimeter approach, start by locking up everything that can be locked up.

LAN hubs

Most LAN hubs are in wiring closets. An amazing number of these closets have no locks on the door. If they do, they are simple snap locks, easily beaten with a credit card or a hefty tug. Worse, some have locks but are left unlocked. This is no help. Sometimes LAN hubs are in an open area. These should be relocated to a secure location. Failing that, enclose them in a mesh wire cage.

Access to the hub means that an adversary can smash the hub to bits. They can rearrange all the ports just for sport. Inserting a shorting jumper into a port can short out the LAN cable and may take days to find. Intruders can tap into a segment where they do not belong. With all the wires around, one more will never be noticed.

There is a strong trend to concentrate control at hubs, adding repeaters, bridges, and even servers into the hub. As more functionality migrates to the hub, the vulnerability of the LAN increases with it *if it is not locked up.*

LAN wiring

The classic tactic with LAN wiring is to hide it in the ceiling, in the walls, or (using flat wire) under the carpet. Granting that cables are unsightly, that fact gives wiretappers a point of entry. They can tap in and the tap will never be noticed. The logical conclusion is that all cables should be exposed for everyone to see. That *sounds* great but *looks* awful. However, if your application is extraordinarily sensitive, then aesthetics may have to take second place.

Difficult as it is, try to be alert to new wiring, wiring of a different type, and wiring that looks unfamiliar or out of place. Be aware of the workstations around the office and do not leave unused workstations operational.

Document all wiring. This should be done anyway, but it gives you a reference. If it is not in the documentation, what is it doing there? Similarly, disconnect unused lobes from the hub. Do not give an adversary an easy entry point through someone else's laziness or neglect.

Unshielded twisted pair cable is the most vulnerable wiring because it is the easiest to work with, and it is everywhere. Coaxial cable LANs are slightly less susceptible. Ethernet 10Base5 LANs require a tap and a medium attachment unit, while 10Base2 systems are easier to tap, needing only a BNC T connector for access.

If you really want to be safe, buy fluorescent orange and yellow cable and put it in plain view. An alternative might be metal or plastic conduit, also in plain view. Taps penetrating conduit will be pretty obvious.

It helps to scan the cable periodically with a cable scanner.

Record the readings after installation. Then you will have a reference. If a tap is added later, it will likely show different numbers.

Unshielded LAN cables emit electromagnetic radiation. An inductive (coil) pickup can sense these signals without a physical tap. LANs using IBM type 3 cable should have media filters to prevent excessive (and illegal) radiation.

Fiber optic cable is difficult but not impossible to tap. There are systems available that detect even small amounts of light loss and sound an alarm. Cordant Federal Systems is one such vendor. Remember that if an optical signal passes through a repeater, then the signal becomes electrical at that point and can be tapped without exotic optical taps.

Radio-frequency and infrared LANs do not use wires. Instead they use radio frequencies or heat waves. Either can be readily tapped. No wireless LAN should be purchased without the vendor's encryption option. All wireless LAN vendors offer it. Do not go broadcasting your confidential information out the window.

What do adversaries listen with?

An ordinary workstation will do. Some workstations (especially if Unix-based) can be set to *promiscuous mode* and will display all cable traffic regardless of destination address. Once attackers have physical access, they can attempt logical access by fishing for user IDs and passwords.

Attackers do not even have to do that. A few hundred dollars' of software will convert a workstation into a LAN traffic monitor. These monitors can decode Novell IPX/SPX packets, TCP/IP packets, or AppleTalk data with ease. They conveniently translate binary into readable English. Some will record traffic for later analysis at leisure.

Monitors are passive: They only read data, and do not create any. Monitor devices are therefore undetectable. Active test devices can create packet messages, making "spoofing," the opportunity for pretense, possible. There is a slow movement toward having all monitors emit presence frames periodically to let a manager know that a monitor is connected. Although the aforementioned movement is glacial, you should inquire from any prospective vendor as to the ability of their monitor to emit such frames.

LAN analyzers may be passive, active, or both. All LAN analyzer software should be tightly controlled. LAN analyzers such as Network General Sniffers and Novell LANterns and LANalyzers should be locked up when not in use. Some offer password protection before they can be used. If present, this feature should be invoked.

Cable breaks

It is estimated that 70 to 80 percent of all LAN problems happen at the physical layer. Of this percentage, many physical problems are related to cables.

Token-ring LANs have inherent backup between multistation access units (MSAUs). If one cable is broken, the ring will automatically loop and restore itself. FDDI works the same way. Clearly, a backup ring connection can work only if it is installed—its use is optional, but widely implemented.

If possible, try to use different risers to give cables an alternate path. A fire in a riser or on a floor will be less likely to take out all the cabling. Access doors and inspection panels for risers should of course be locked.

Many Ethernet hub vendors offer redundant hub cable schemes. They are based on redundant star, mesh, or ring cable topologies to create a duplicate path should a main link fail. Keep in mind that wireless LANs offer an emergency backup if all else fails.

Servers

All servers should be under lock and key. Too often they are left out in the open. This gives easy access to the system console and especially to the floppy disk drive into which a virus-contaminated disk may easily be inserted—or anything else for that matter.

Aside from accidentally spilling coffee on it, the server can be knocked over or otherwise come to harm. A server once running does not need much attention and tends to become invisible. This is good and bad.

A case that is familiar to the author illustrates the point. A server was found to have a corrupted disk drive. It was replaced. The problems continued. Eventually the entire server was replaced. There was no improvement. In desperation, a staff member (probably junior) was delegated to watch the server overnight from a secret place. About 1:30 in the morning, the security guard entered the room, sat down, and clamped his magnetic flashlight to the side of the server and fell asleep. Problem solved.

Novell in particular is emphatic about locking the server up. Breaches of NetWare security have occurred because someone had access to the server console itself, and we will discuss a case example of just such an occurrence in the next chapter.

Electrical security

It is said that we are running twentieth-century electronics with nineteenth-century electricity. That is true. Servers should be equipped with uninterruptible power supplies (UPSs) that also condition the power so that servers always get noise-free and stable electricity. In case of a power failure, the power supply should inform the server so the server can perform an orderly shutdown. Otherwise, damage to data and application files can result, making data loss probable even if backups have been done. In extreme cases, physical damage to the server can result, such as a disk head crash or an overheated power supply (which can occur during a "brownout" low-voltage condition).

There are many documented cases of power conditioners and UPSs that saved the day. The two are not the same: power conditioners clean up power while the UPS adds short-term power availability during the server's shutdown cycle. Additional critical devices, such as hubs or concentrators, bridges, and routers may also be candidates for power protection.

Grounding

Good grounding practices help prevent intermittent and flaky operation. Poor grounding can even cause chip damage due to static electricity discharges. Poor grounds stem from two sources. Sometimes the third wire in an electrical plug is missing. Sometimes the third wire ground on an outlet is not really connected to ground or is poorly or improperly grounded.

Another common problem is that the wires on the outlet are reversed. There are two wires, usually black and white. The black wire is electrically "hot" and should go to the left-hand pin.

You should not feel a vibrating sensation when lightly rubbing a metal case. If you do, it is electrically "hot." Reversing the plug or wiring solves the problem. In one case, a server that crashed repeatedly for no apparent reason was found to be floating some 60 V above ground. Workers got a tingle every time they touched it, but mistakenly thought it was static. Be safe. You do not want to be defibrillated accidentally.

Static

Servers, workstations, monitors, *everything,* should be grounded. Static mats, static spray for carpets, and grounding straps or touch pads are advisable for those working on computer internals. It only takes a 2-kV static discharge that you cannot even feel to damage a chip or a circuit board. Damage may be partial, complete, or at worst, intermittent.

Ethernet–IEEE 802.3 Protocol Hub Security Features

One of the easiest ways to add security to an Ethernet LAN is to do it through the hub. The concept is simplicity itself and is shown in Fig. 7.2. Each network interface card has an address. The card creates Ethernet–IEEE 802.3 frames containing both a *destination address,* indicating where the frame is going, and a *source address,* showing where the frame is coming from.

Ordinarily, frames entering the hub are broadcast to all other stations. This creates a great security exposure and violates the well-known "need-to-know" principle: If they do not need to know, do not tell them!

A hub equipped with address security contains a table that relates the frame's destination address to the hub's destination port number. The incoming frame's destination address is examined and sent to that specific port. All other ports receive random

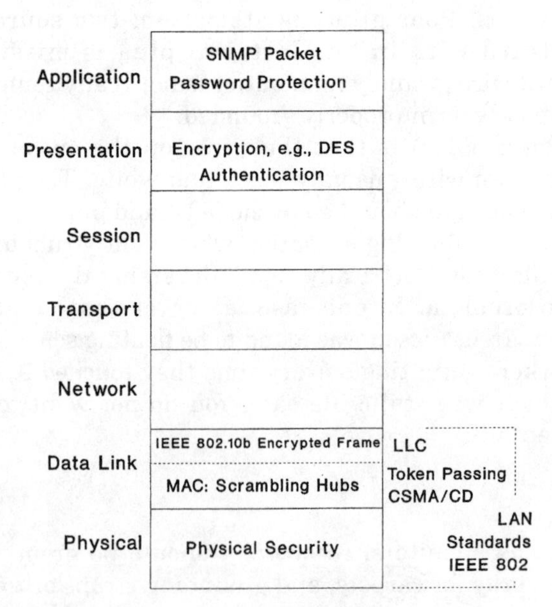

Figure 7.2 Hub address verification security.

1s and 0s. This is not encrypted data, just 1s and 0s sent to maintain timing and for traffic security.

Some hubs also examine the incoming source address and relate it to a specific port number as well. The hub will not accept the traffic if there is a mismatch and will notify the manager if a management system is in place. The system may be used to create *closed user groups* by permitting a source address to communicate with a limited number of destination addresses. For instance, a station might be allowed to connect to one server but not another. Multicasts and broadcasts can be accommodated by examining the address to see if the frame is of either type.

People who move their machines or who try to connect to an unused port are automatically locked out. An NIC address not recognized by the hub will also be denied admittance, keeping out foreign machines. Depending on the vendor, some or all such events can set off alarms. Some systems allow the manager to lock out ports or make logical port and node connections remotely, with or without security.

Token-ring hub security

Since the token must be passed from station to station in sequence, the port/in to port/out technique cannot be used. However, a frame entering the hub can be examined for source address relative to port number validity. The hub port number as previously related to a specified destination address can also be verified. Port-to-address validity checking as a security technique is equally valid for FDDI networks.

Discussion

Hub-based security is a physical layer approach. Such hardware security has great advantages:

- It is fast. Hardware chips such as application-specific integrated circuits (ASICs), SONIC, and reduced instruction-set computers (RISC) chips are all used. For instance, Hewlett-Packard uses an Intel i960 RISC chip.
- Minimal delay is important. Ethernet imposes strict time delay limits on Ethernet repeaters, which is what these hubs basically are.
- Hardware-based silicon implementations are hard to modify and thus compromise.

- Hub-based security is protocol independent. Upper-layer protocols have no effect. Interoperability troubles do not exist.
- Hardware security is preferable to filtering via bridging, for reasons of speed and administrative convenience.
- At about $60 per port, hub-based security is inexpensive, typically 10 percent of the cost of DES encryption. Often it can be installed in the field, too.
- Secure hubs are exportable, since no encryption is used.
- Security can be implemented at some workgroups and not others. It is a very scalable solution.
- The scheme works with any Ethernet–IEEE 802.3 vendors' cards.
- Setup and administration are minimal.
- Most auditors will be satisfied with hub-based security provisions.

Terminal servers

One very good way of providing an additional access control perimeter is to use a terminal server as shown in Fig. 7.3. A terminal server is placed between the LAN cable and a dumb terminal

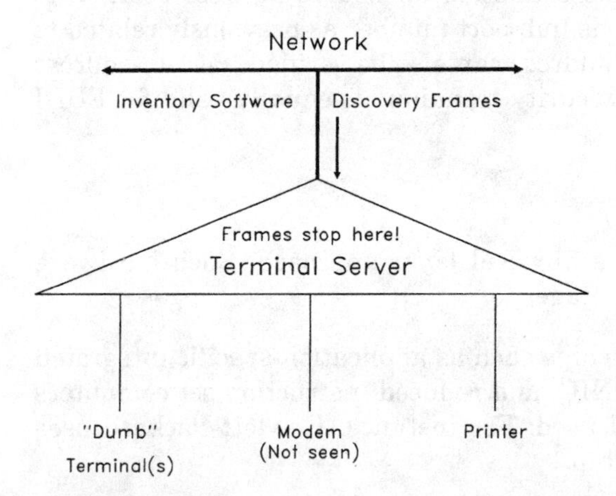

Figure 7.3 A terminal server.

or a PC running a terminal emulation program. Usually there are eight or more such connections possible. Printers and modems can also be connected to some terminal servers.

One must log into the terminal server before LAN access can be gained. The cable between the terminal and server carries only information for that terminal or device, an added security benefit.

Dial-In Communications

It is easy to break LAN security through the dial-in route. Without management, a caller can dial in and try passwords until they gain access. Once in, they can hide pieces of software anywhere, pass through WAN links to other systems, and generally create as much or as little havoc as they like.

Remote users with notebooks and laptops should never store their passwords in plaintext login scripts. These should always be entered by hand. Alternatively, the password protection in some notebooks should be invoked in case a notebook is stolen. Be advised, though, that passwords stored in CMOS memory, as with some Toshiba, Compaq, and AST computers, can be bypassed if the CMOS battery is removed.

You may use a software lock utility, such as SECUREcard/110 from DataMedia (Nashua, N.H.), DiskLock PB from Fifth Generation Systems, or Norton Essentials from Symantec. You can also encrypt sensitive files in notebooks using Guardian Plus (Command Software Systems) or CA-Cortana from Computer Associates (Islandia, N.Y.). Macintosh PowerBook users can use Datawatch's (Research Triangle Park, N.C.) Citadel or the Macintosh version of DiskLock. All these programs require a password before decryption can occur.

For fixed machines, Trend Micro Devices [Torrance, Calif.; (800) 228-5651] offers an 8-bit ISA-compatible circuit board that checks for viruslike activity and restricts user access via password. To prevent bypassing security by removing the board, StationLock encrypts the PC's partition table. This makes booting and hard disk access impossible.

The conventional response to the random dial cracker has been the dial-back modem. When a call is answered by the modem, the caller enters a code. The modem looks up a corresponding phone number and calls it back.

Several things can go wrong. For one thing, a hacker can record the caller's code and use it later if she or he can be at the target location, turn on call forwarding, or otherwise divert the call. Some dial-back modems do not listen for dial tone when they call back. If the cracker stays on the line, dialing occurs but has no effect. Then the cracker starts trying passwords until they achieve entry.

Solutions vary. Information Resource Engineering (Baltimore, Md.) offers a packet-sized authenticating and encrypting modem with random password generation by call. Authenticated by a user's PIN, the modem calls a companion device at the central site. Once the two devices shake hands to perform the authentication process, the caller has legitimate access. The use of encryption during the call is optional.

National Semiconductor sells PersonaCard, a PCMCIA electronic identity card that uses RSA public keys. Messages created by the user can be electronically signed with their private key. Other PersonaCard holders with the user's public key can verify the sender's identity. Up to 30 public keys can be stored in the tamper-proof PCMCIA card, with a larger version that can store 100 keys. The price is approximately $250 for the former and $279 for the latter. The card can also encrypt data.

Security Dynamics (Cambridge, Mass.) offers a token-based password system called SecurID that resides between a modem pool (connected to a server or mainframe; no difference) and the telephone line. A caller enters a PIN on a $50 card that fits into a wallet or purse. Then they enter the number shown on it into the system. The central site password changes periodically, every 30 seconds or so. If the passwords match, the caller is granted access. Note that this is purely an authentication system and that there is nothing to stop a hacker from kicking the legitimate user out once the exchange is complete.

SecurID can also be used with Apple Computer's Remote Access Multiport Server. Working in conjunction with IBM, OS/2 support is also planned.

IBM offers NetSP, its Network Security Program. Under NetSP, a client requests access to a application controlled by NetSP. NetSP authenticates the user and grants the user a security ticket. The server (or mainframe) receives the ticket from the NetSP authentication server and opens the requested application. NetSP can accept a PIN or password from Security Dynamics' SecurID system instead of a NetSP password. This combines user authentication with the single-sign-on feature of NetSP and facilitates remote users. NetSP is unique in that it offers access control for both server and mainframe applications in a Kerberos-like security architecture.

Paralon Technologies sells a hardware-based solution using a pair of matched PathKey devices at each end of a connection. The

PathKeys generate passwords and encryption for each session. One PathKey can recognize up to 500 other PathKeys; they simply have to be directly connected for a few seconds before deployment. Simple to use, which is an essential for good security, PathKeys sell for $280 each.

A list of hardware-based security vendors is found at the end of the chapter.

Configuration control

When the workstation is booted, commands can be loaded into the configuration files such as AUTOEXEC.BAT to connect to a specific server or service at power-up. This gives the user no options and less opportunity to browse.

Encryption

Encryption is one of the less desirable options in a LAN. While it offers a very high level of security, it is expensive whether implemented in software or hardware. There is also a performance penalty. Consequently, one must be really sure the application calls for such protection before proceeding.

Software encryption exacts a heavy toll on performance, reducing the throughput of an average network interface card from 400 kbyte/s or so to around 320 kbyte/s or 80 percent of its normal throughput. Software encryption is an option only when this type of delay can be tolerated. While short messages may see little effect, graphic and Windows applications will suffer.

Hardware encryption is faster, but still reduces NIC throughput to around 90 percent of its normal value. The hardware is expensive, roughly a third the cost of a PC. Another consideration is the key management problem. Key management adds a significant administrative burden.

Partial encryption can also be used under the IEEE 802.10b protocol that resides at the logical link control sublayer. Shown in Fig. 7.4, the outer header (called an "envelope") is not encrypted and contains addressing and handling information.

The inner envelope contains security-related information. Upper-layer information such as IP datagram addresses and TCP data plus user information is encrypted. This makes the message just like an envelope for a letter. You can read the address, but not the contents.

The standard defines the types of encryption keys that can be used, but not the encryption scheme, such as DES. The IEEE pre-

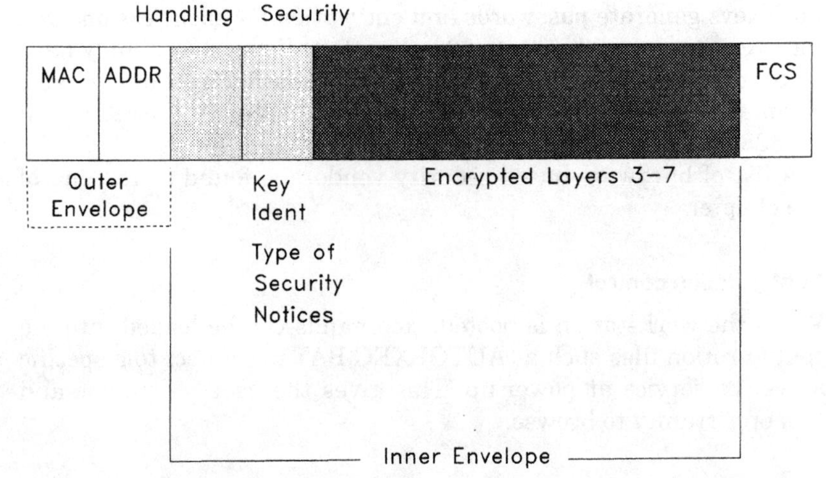

Figure 7.4 The proposed IEEE 802.10b security packet.

ferred to leave that decision to the users. In a security association such as that between a client and a server, the IEEE 802.10b protocol also prescribes the types of security notices that must occur.

As yet, this standard is still new, so few products exist. As such, performance and cost issues remain open. The IEEE 802.10b protocol can complement the hub security scheme described earlier.

Authentication

Authentication may be used with LANs. As discussed in Chapter 4, hashing is the usual method, attaching an encrypted message digest. The NIST versions of MD2 and MD5 hashing algorithms are typically used.

Message digests are usually built using software. Even so, they have little effect on throughput, with speeds possible up to 2 Mbyte/s. Ethernet operates at 1.25 Mbyte/s and IEEE 802.5 at 0.5 and 2.0 Mbyte/s for 4 and 16 Mbit/s token rings, respectively. Accordingly, MD2 and MD5 can keep up with these popular LAN technologies.

Bridging

As shown in Fig. 7.5, it is possible to create closed user groups with a bridge. A frame received at the bridge is examined to see if the sender and recipient are in the same user group. If they are, the frame is passed through the bridge. Otherwise, it is discarded. In a way, the bridge is performing the same source–destination comparison as the port-in–port-out technique.

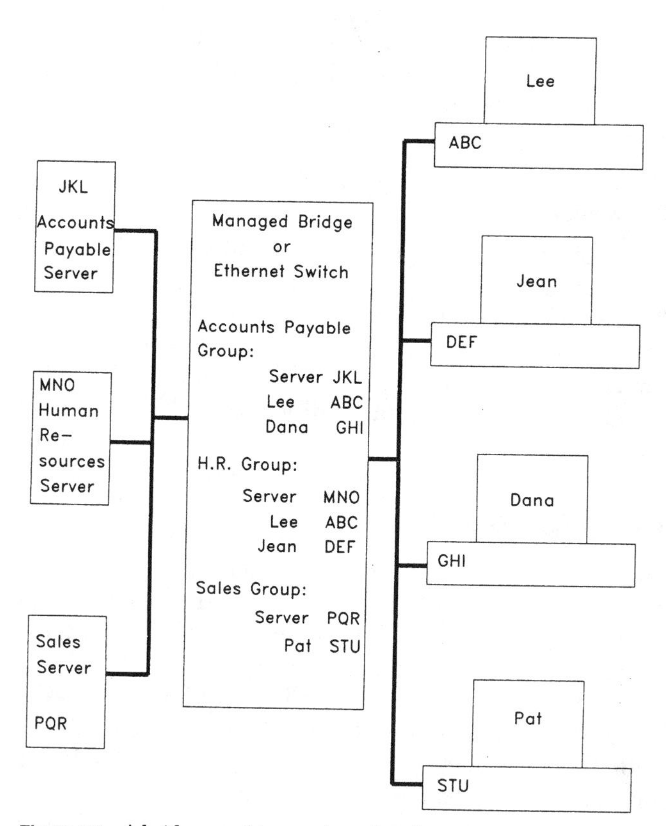

Figure 7.5 A bridge used to create a closed user group.

Bridge filtering and forwarding is a performance enhancer, since unnecessary traffic is filtered out and so does not appear on the second LAN segment. It is easy to do a second check of the source–destination address to decide whether the source should have access to the destination and, for that matter, whether the source should have access to any destination beyond its own LAN segment.

Bridges do exact a small performance penalty, but they work with all LANs, including Ethernet, token-ring, and FDDI local area networks. Another virtue of a bridge is that it can find an alternate route using the IEEE 802.1d Spanning Tree Protocol should a link fail.

Power failures in hubs, bridges, and routers

Hubs or concentrators, especially those equipped with internal server or router cards, may have the same orderly shutdown requirements as standalone servers. Accordingly, it is becoming common to provide redundant power supplies in hubs, both supported by a UPS. Sometimes power conditioning is added, usually as part of the UPS.

Electromagnetic radiation

Computer screens emit strong signals, about 40 times the level emitted by cables. These signals can be read if someone wishes to try. If successful, the intruder has bypassed all your security.

One preventive option is to use LCD screens, whose radiation level is very low. A more expensive option is to use shielded (TEMPEST) CRTs. Very few sites demand a fully shielded TEMPEST-secured building.

Security Risks in Using SNMP Version 1

The Simple Network Management Protocol is the universal protocol of choice for LAN management. It is widely implemented to manage everything from hubs to servers, workstations, bridges, and routers and is a product of the Internet community.

SNMP is based upon five basic commands. Four are read-only, but the set command orders machines to do things. The lack of adequate security for the set command led many vendors to disable it or simply not implement the command.

For instance, it is possible in SNMPv1 to capture a set command containing the SNMP community string and then edit and play a modified packet. This tactic could be used to disable a router or turn off the hub hardware security we have just discussed.

SNMPv1 provides no confidentiality and only basic authentication that does not even encrypt the SNMP-based password. In response to this deficiency, the Internet Engineering Steering Group (IESG) asked for corrective proposals in March, 1992. The IESG identified four key areas that needed to be addressed:

- *Origin authentication:* to prevent masquerading by a spoofer.
- *Replay protection:* to prevent recording and reissuing a packet at a later time.
- *Message integrity:* to prevent modifying a packet.
- *Data encryption:* to permit use of DES to encrypt SNMP data.

SNMP version 2 was the result of the IESG's request for proposals.

SNMPv2

SNMPv2 uses a concept called *parties*. Managers and agents (see Fig. 7.6) contain one or more parties. If the sending party includes authentication, the receiving party will read it. If the receiving party asks for privacy, the sending party will encrypt the message.

Access controls can be imposed on an agent's management information base (MIB) to make all or part of it available to others and limit the operations that a remote party can perform on it. As a result the parties chosen, be they manager and agent or manager and manager, mutually decide on the security features they need. As Fig. 7.7 shows, in essence a three-dimensional matrix consisting of managers, agents, and functions allows the network manager to create any desired level of access control, authentication, and privacy (encryption).

To do this, the manager uses a party MIB implemented for each manager and agent. The MIB lets the user configure the desired

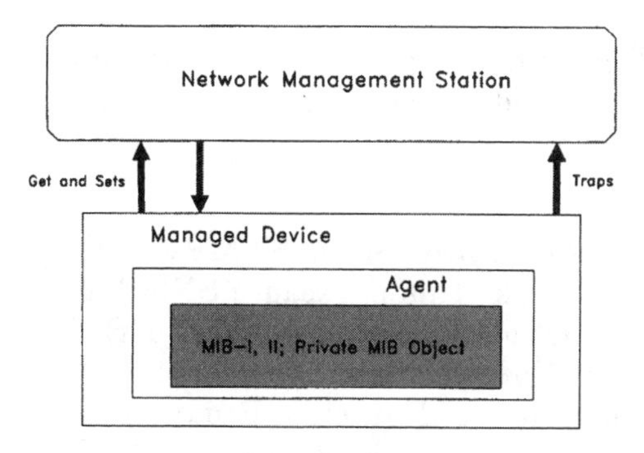

SNMP Proxy Agents allow non-SNMP devices to be managed.

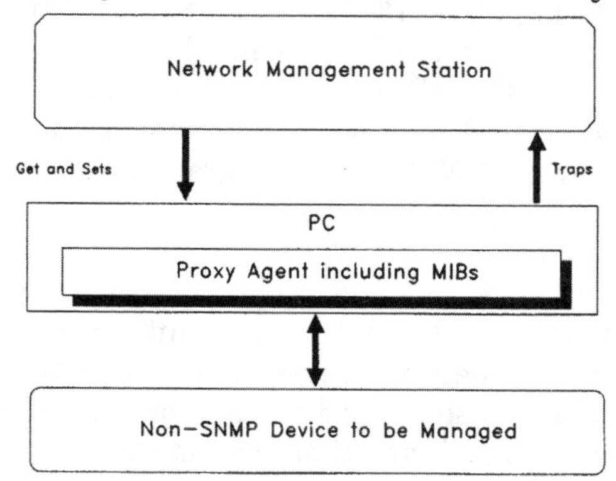

Figure 7.6 An SNMP manager and agent.

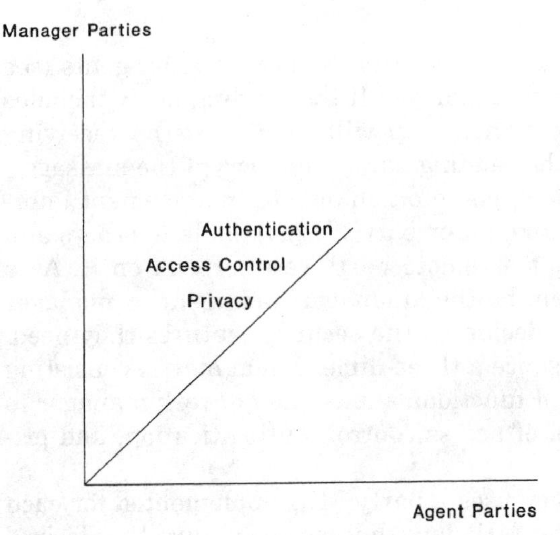

Figure 7.7 The SNMPv2 security matrix.

security level for each local and identified remote party, accounting for all possible relationships and security features.

Compatibility with SNMPv1

SNMPv2 is backward compatible with SNMPv1, although SNMPv2 features will not work. This is because SNMPv2's structure of management information (SMI) is a superset of SNMPv1's. All MIBs defined for version1 are usable with version 2.

SNMPv2 has been implemented quickly. Products began to appear late in 1993 once SNMPv2 was adopted by the Internet Engineering Steering Group. The IESG, while not a standards-making body such as the ISO or IEEE, does handle the standardization of protocols based on TCP/IP, which includes SNMP. (SNMP relies upon the User Datagram Protocol for end-to-end communication and upon IP for routing of SNMP user datagrams).

Inventory Management

Inventory management as an activity has a definite role in network security. First, it deters theft when people know equipment is tracked. Second, it discourages self-styled power users from reconfiguring, adding, deleting, or modifying their workstations.

Perhaps the greatest appeal of inventory management is that by using discovery software, managers can get a comprehensive view of what is *really* hooked up and how, as opposed to what they *think* is out there. The *discovery function* permits a network manager to discover all connections to the network whether documented or not. Managing inventory makes the management of individual workstation configurations at least conceivable. The SNMP protocol and intelligent hubs allow central control over a decentralized network. Inventory management consolidates that control.

International accounting firm Coopers and Lybrand has LANs in each of their 60 offices across the United States. The entire SNMP-hub-based network is managed remotely from its headquarters in New York City, including moves and changes in each office. This company has shown that central control *is* feasible.

Inventory management software

Like any software package, the "personality" of inventory management software is shaped by the philosophy of its designers. Some designers seek to gather inventory and no more. Others take an accounting approach and see the software as an asset-management tool. Another type of software designer takes the viewpoint of the network manager whose main interest is desktop resource changes. The last approach is that of vendors like Hewlett-Packard who see inventory management as an element of overall network management.

Indeed, the OSI specific management functional area (SMFA) model includes inventory management as one of its core elements. Hewlett-Packard's powerful Windows-based OpenView network management system is a fully SNMP-compliant system that can manage all the devices we have mentioned so far.

Hewlett-Packard offers a pair of software packages for performing network inventories. The first is OpenView Workgroup Manager, which can run on a 386-based PC. It lets managers monitor and control devices in multivendor workgroup networks and automates many ordinary management chores.

As a discovery tool, Workgroup Manager automatically finds all devices on the workgroup LAN, including hubs, printers, PCs, Novell NetWare servers, and clients. It then builds a map showing the actual topology. The discovery process can be programmed to take place periodically to show additions, deletions, and changes.

As a multivendor tool, Workgroup Manager can look at SNMP-managed or Novell Internetwork Packet Exchange (IPX) devices. MIBs from Hewlett-Packard or others can be browsed and MIB counter contents shown graphically. The tool costs about $2,000. Interconnect Manager goes beyond the workgroup to do the same for interconnected workgroup LANs. Because of its greater capability, this version is $4,000.

Other packages with a network-management viewpoint include:

LAN Workstation by Saber Software

LANDesk by Intel

LANlord by Microcom

LANExam by Network Computing

Evaluating inventory software

First decide on your needs. If your need is asset management, look for a tool that stresses that. But for any tool, evaluate the following:

Inventory accuracy. The software should evaluate installed hardware such as disks and memory accurately. Software accuracy is best determined using the CRC authentication technique discussed in Chapter 3.

Depth of audit. How deep can the tool go? Down to the NIC card? You should be able to select the depth of analysis needed.

Reporting ability. An ad hoc report generator should be an integral part of the package. "Canned" reports alone are not adequate. The tool should have an export feature so that any report generator or data reduction tool can be used in conjunction with it.

Alerts. Alerts warn of possible theft and can help in troubleshooting and also asset tracking. These amount to reports by exception.

Automation. The tool should contain a calendar to set audit times. To the extent possible, installation for the tool should be automatic.

Ease of use. If it is not easy to use, no one will use it, no matter how it excels in the above categories.

A list of desktop inventory software packages is located at the end of this chapter.

Summary

We have looked at LAN security from the vantage point of hardware and network management. Into this we have drawn most of the LAN security threats shown in Fig. 7.8. In the next chapter, we look at the last remaining but vital LAN security item, the network operating system.

Also, see NIST Publication FIPS 191, *Guidelines for the Analysis of LAN Security.*

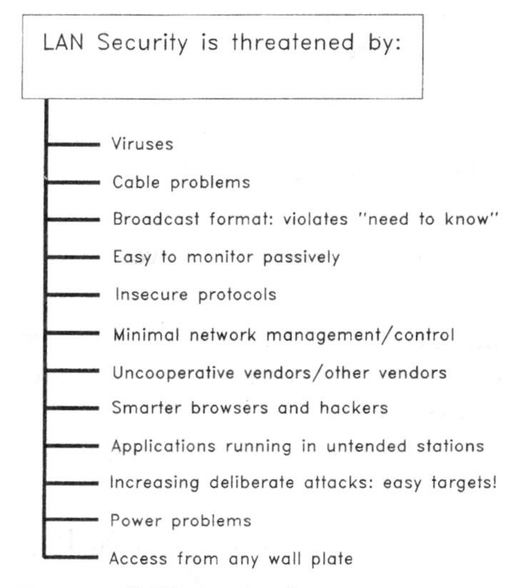

Figure 7.8 LAN security threats.

Review

- LANs are vulnerable because they are widespread and decentralized.
- Using the perimeter approach, lock up everything: hubs, servers, bridges, etc.
- Lock up wiring closets and riser inspection ports, and use several wiring routes.
- Security must include recovery from an accidental or deliberate cable break.
- LAN analyzer software and hardware can eavesdrop on and spoof workstations easily; some workstations can do so natively as a built-in feature.
- Electrical security includes uninterruptible power and/or power conditioning.
- Grounding is not only a network security hazard but a personnel hazard.
- Static electricity can be managed with conductive mats, touch pads, and sprays.
- LANs violate the need-to-know rule by broadcasting everything to everyone on the same LAN segment, be it a bus or a ring.

- Comparing known valid addresses to hub ports is a very good way to provide physical security in LANs. It works for Ethernet, token ring, and FDDI.
- Ethernet messages can be sent directly from port to port; all other stations get gibberish. The need-to-know security rule is thereby satisfied.
- Hub security is fast, inexpensive, interoperable, and works with NICs from any vendor.
- Terminal servers eliminate the need-to-know problem. They also add another password layer between the terminal server and workstation before network access is allowed.
- Limit access by presetting configuration files to discourage browsing.
- Encryption in any form is a marginal option for a LAN: It is expensive and slow.
- Authentication is a better choice, provided that it is sufficiently secure for the need.
- SNMPv2 overcomes SNMPv1's security limitations very flexibly using parties.
- Inventory management, especially network discovery, enhances security.

Appendix. Desktop Inventory Software and Hardware-Based Security Vendors

Desktop inventory software

Frye LAN Directory

Horizons LANauditor

Intel LANDesk Manager

McAfee Associates Brightworks

Network Computing Bindview

Saber LAN Workstation

Symantec Norton Administrator

Tally NetCensus

Visisoft VisiNet

See a review of the above packages in *Network Computing*, September, 1994, p. 130.

Hardware-based security vendors

CE Infosys
Herndon, Va.

CRYPTOCard Inc.
Palatine, Ill.

Cylink Corp.
Sunnyvale, Calif.

Digital Pathways, Inc.
Mountain View, Calif.

Enigma Logic, Inc.
Concord, Calif.

Fischer International Systems Corp.
Naples, Fla.

Information Resource Engineering, Inc.
Baltimore, Md.

Mobius Encryption Technologies
Washington, D.C.

Secure Computing Corp.
Roseville, Minn.

Security Dynamics, Inc.
Cambridge, Mass.

Semaphore Communications Corp.
Santa Clara, Calif.

UsrEZ Software, Inc.
Seattle, Wash.

Chapter

8

Finding Network Operating System and Operating System Weaknesses

"What is driving security on LANs is the value of information that is now held on them."

TODD DAGRES, *Consultant*

Objectives

At the end of this chapter you will be able to:

- Specify ways to make your server more secure
- Use network operating system (NOS) audit tools to keep tabs on server activity
- Describe methods used to secure workstations
- List ways to secure computer hardware physically
- Discuss the relationship between security and disaster recovery
- Understand how directory and file attributes affect security
- Explain security tools present in NetWare 3.X and 4.X
- Discuss ways to protect NetWare servers from viruses
- Describe ways to provide workstation virus security
- Compare and contrast workstation and server-based virus scanning techniques

- Identify where the virus' back door is and how to close the door
- Explain the University of Leiden NOS type of breach and how to prevent it
- Discuss the Riverbend Group type of breach and how to prevent it from happening to you
- Dissect the weak spots in Banyan Virtual Network Services (VINES) and Microsoft LAN Manager NOS
- Identify attack techniques in operating systems such as time of check to time of use (TOCTTOU), "Superzapping," "browsing," and the "salami attack."

One of the most common flaws in network security is taking things for granted: The server has never failed. Workstations work. Life goes on.

Right around noontime one day in February, 1993 there was a major attitude adjustment for a lot of people. They were employed in the World Trade Center.

In the days following the bombing, a few people were let into the dark, cold, and soot-encrusted building to retrieve precious computers, disks, tapes—whatever nervous administrators could get out on a handtruck. The television coverage showed PCs, terminals, and every kind of equipment emerging from the dark tower.

How much of this equipment remained usable? The data recovery companies made fortunes trying to recover data from disks that were black with soot. Unsuccessful recoveries cost companies money, information was lost forever, and more than a few administrators lost their jobs.

LAN proponents learned a bitter lesson that day, one their mainframe cousins had learned long ago. The lesson had to do with keeping backups offsite and not allowing one's practices to get sloppy because "nothing ever happens." Contributing to the mistake was the perception that because the server was small and did not cost much, it was nothing special. Wrong. It was the invisible value of the information *inside* that made it worth protecting. Such data may be irreplaceable or can be rebuilt only at great expense.

It does not take a bomb. It only takes a virus. Perhaps it takes a hacker with a LAN analyzer and a grudge to settle.

Contents of This Chapter

We begin with some recommendations for all systems, regardless of NOS or OS. From there, we become more specific with respect to

NetWare, VINES, Apple's System 7, and LAN Manager. Even specific system flaws may have relevance to other NOSs, so it is wise to pay attention to all of the text even if you do not have the particular OS or NOS under discussion.

We also discuss some operating system holes in DOS, Unix, and OS/2. Microsoft's security plans for the Windows NT operating system will also be reviewed. In addition, there have been two significant NOS breaches that have lessons for all NOS users, and we will discuss them both.

Recommendations for All Systems

Servers

Disable the server's floppy disk drives. Either unplug them or make them otherwise unusable. This way, hacking programs cannot get into the server nor can a boot sector virus.

- If an adversary brings the server down, he or she cannot boot it as a DOS machine and cannot copy any files.
- NetWare users can enter a console command, REMOVE DOS, to do the same thing.
- Use a screen blanker with a password to lock the console. In this way, a malicious program loaded into a user's directory cannot be transferred to the server from the console.

Network World, the trade paper, has a bulletin board with a number of security programs on it.[1] NLMLOCK2 is their screen blanking program.

NOS security attributes

Use them all. It is amazing how few security features are actually used, even though users have paid for them. A prime example is NetWare's Intruder Alert, which notes repetitive attempts to login. It does no good if disabled. Use NetWare's Security utility or other tools such as Network Security Organizer to discover network weak spots.

Review the audit trail every day. NetWare 3.X does not have audit capability, so use an NLM or Network World's PAUDIT2. Add it to your login script to be sure you do it.

In the server's directory tree, deny users write permission in directories where executable files are stored. *Directory-level* denial is necessary as most viruses can get past the file-level read-only attribute.

Workstations

Network supervisors should log in from a limited-access workstation so that it cannot become infected by a virus. Disable (electrically disconnect) drive A on workstations. Boot viruses get to the hard disk from drive A. B drives are not an entry point.

Virus-scanning network interface cards are available from Intel. This is an ideal place to detect and stop a virus.

User login scripts should use the DOS COMP(ARE) command to compare COMMAND.COM in the workstation to a reference copy in the server. Use the same script to do a local virus scan and limit the connection(s) allowed (if possible).

The DOS MEM(ORY) command shows the workstation's memory. In the script, redirect the information to a file. The next command line should invoke a program to read the file to see if everything is all right. If not, the program should halt the login.

If the workstation can be physically locked, lock it. If it can be protected by password, protect it. Programs such as PROT and PASSWRD are on the Network World Bulletin Board System.

Get physical

Here are some physical security suggestions:

- Spray computer gear with odd colors to reduce its fence value after theft.
- Cover one screw with nail polish or auto paint so that if someone tries to open the case, the seal will be broken.
- As noted before, lock all servers in a secure room. Limit access and keep keys in a key safe or other secure area.

Security guards should have set procedures for allowing computer gear out of the building. Passes are not enough. They should have a list of who can and cannot take gear out. Some experts recommend photographing anyone removing equipment.

Record equipment numbers in a log, perhaps within the inventory management system. Nonserialized units should be stamped or engraved with your organization's name or equivalent. Nonremovable labels can also be used, especially inside units where thieves rarely look.

Your insurance agent may be able to get you an IDENTIFAX number. This number is stenciled on everything of value. Hardware can be traced easily by police through the IDENTIFAX registry.

Another topic to discuss with the agent is what proof is required to make a claim and the terms under which computer gear is covered. Receipts and periodic inventories may be necessary to demonstrate ownership and theft if it occurs. Specialized policies exist to protect against damage or theft.

When the worst happens

Despite our best efforts, sometimes the worst happens. Note the World Trade Center bombing, or the massive failure at the Hinsdale, Ill. central office that caught fire. For these reasons, the security plan must be inextricably tied to the disaster recovery plan. The center of the disaster recovery plan is backup. Back up everything to a tape that has more capacity than the server drives. Back up only material you really need. Delete duplicate server files.

Network users should be able to back up to the server via menu. Insist that they do so every day. Files not needed by the user can be moved to tape and deleted from the server.

Figure 8.1 illustrates the concept of *online, nearline,* and *offline* storage. Utilities exist that automatically move data downward after it has been unused for a period set by the manager.

The author does not bother to back up .COM or .EXE programs unless they have been substantially customized. These can always be restored from original disks or duplicate originals. Sometimes its possible to back up only the amended files.

In setting backup parameters, invoke as much error checking as possible. Speed is less important than reliability. Laptop users are not excluded from their backup responsibilities. Laptop users can use external "zero-slot" or PCMCIA network interface adapters to back up their disks to the server.

Test your ability to restore now and then. When the real thing happens, despite your nervousness, you will not fail.

Novell NetWare

NetWare has excellent security. Security features were built into NetWare from the beginning, not added on. NetWare's special file structure exists for performance reasons, but has the additional

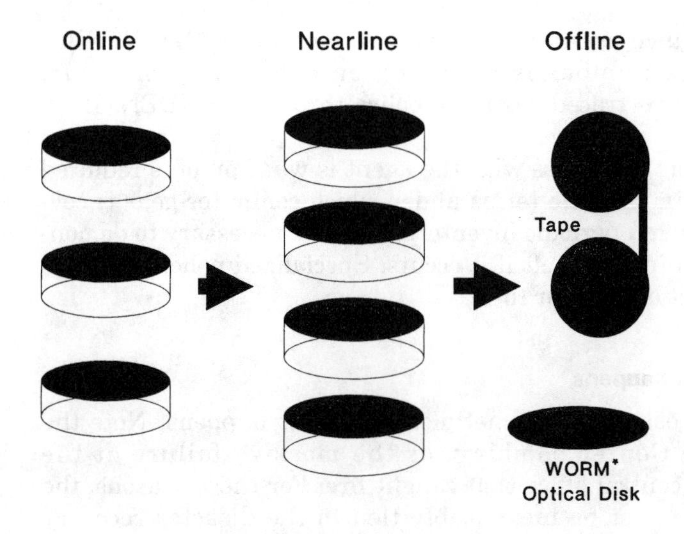

Figure 8.1 Online, nearline, and offline storage.

benefit of not being readable in DOS, Unix, or OS/2 operating systems, even if the user has direct access to the server.

NetWare provides the following:

- Account security
- Password security
- Directory security
- File security
- Internetwork security

User profiles itemize resources to which the user has access, and the rights (Table 8.1) the user has within that resource. The profile can specify the times, dates, and location(s) of access for a specified user.

Directory-level security allows changes to the maximum rights mask (Table 8.2) of a directory. Access attributes can be assigned to directories and files (3.X and up only) to limit access.

Network-level security employs intruder detection and the security monitor.

1. *Intruder detection and lockout* notifies the administrator if excessive access attempts are made. The administrator sets how many incorrect attempts will be allowed and any time delay between successive login attempts. The manager has the option of

TABLE 8.1 NetWare Rights

Trustee Rights*	
Right	Privilege
Read	Open and read files.
Write	Open and write to files.
Create	Create directories and files and write to files. Users cannot reopen files they create without the read right.
Erase	Delete a file, directory, and its subdirectories and files.
Modify	Rename files and directories and change their attributes.
File scan	View directories or files in file searches.
Access control	Modify trustee assignments and maximum rights masks. Grant or revoke other users' rights. Remove trustees.

	Maximum Rights Mask†						
	Read	Write	Create	Erase	Modify	File scan	Access control
Trustee rights	X	X		X	X	X	X
Directory rights	X	X	X			X	
Effective rights‡	X	X				X	

*Trustee rights are rights assigned to users.

†A maximum rights mask is composed of rights assigned to trustees and directories.

‡Users' file rights depend on the trustee and directory rights combined to form effective rights. These plus the directory and file attributes determine the user's rights in a file.

TABLE 8.2 Directory and File Attributes*

Attribute	What it does
Hidden	Hides files from directory (DIR) scans.
Indexed	Creates an index in memory to improve file access; usually reserved for large files.
Read write (normal)	Lets users modify files. Files are normal unless otherwise flagged. They return to normal when other flags are removed.
Read only	Users may read but not modify files. If not set, files are automatically flagged "read write."
Sharable	Allows several users to access files at the same time.
System	Hides files from DIR scans. Prevents deleting or copying. Usually assigned to operating system files.
Transactional	Tracks the transactions in a file. Ensures that all changes are made to a file, or none are. This is to prevent database corruption.

*Directory and file attributes cancel the effect of trustee assignments and rights permitted by the directory's maximum rights mask.

locking the account permanently, or if a specific lockout time is selected, how long that time will be.

2. The *security* utility checks the server bindery (where user records are kept, as Fig. 8.2 shows) for possible breaches. Six breach types are detected, including the following:

- No password associated with a username
- Insecure passwords: too short or not unique to the user
- Users who have supervisor equivalence are noted
- Users with rights to the root directory are noted
- A user who borrows someone else's login script
- Excessive rights in the default directories

Passwords are encrypted both on the hard drive and on the cable.

In NetWare 3.1X, file access rights, encryption services, and secure network console features were all strengthened. However, there was and is no audit trail capability built into NetWare 3.1X. It must be added as a NetWare Loadable Module or other application.

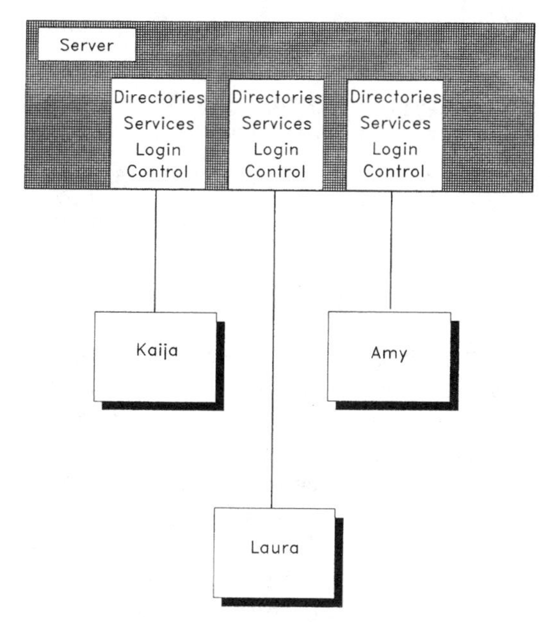

Figure 8.2 The NetWare Bindery.

Automating the assignment and interpretation of rights

Going through each file by hand and trying to interpret the rights level mentally quickly becomes a mind-bending process. Consistent with our stated goal of trying to automate everything possible to gain better security, a number of vendors offer security managers.

One such vendor is Leprechaun Software International, Ltd.[2] Their Network Security Organizer (NSO) guards against viral infections in NetWare servers. NSO resides in the server. It scans the server as users log in, looking for vulnerable directories. Then it creates a checklist indicating the directories at risk and from which users. Rights can then be modified.

NSO provides the following services:

- Analyzes the network's vulnerability to viruses.
- Allows central management and updating of antivirus software in network workstations.
- Performs asset management for workstation hardware and software. A workstation inventory device driver informs the manager of any hardware changes, e.g., RAM or disk size.

- Logs programs and activities from a workstation floppy disk. This feature may help to track down the origin of a virus.
- Logs updated or new programs plus any programs that request write access to another program, as many viruses do.
- Keeps statistics on the number of virus episodes by users and all attached devices.

Workstation controls

All the controls built into NetWare apply to any kind of workstation, whether DOS, Unix, or OS/2. There is a problem with Macintosh computers. Macintosh machines do not give NetWare 3.11's management utilities specific network addresses as do IBM-compatible computers. This means that the administrator cannot assign, restrict, or exclude access to particular network addresses by Macintosh machines. They are effectively invisible to NetWare 3.11.

This quirk has been responsible for a number of NetWare breaches, especially in the academic community. One breach consisted of a Macintosh program that trapped and recorded passwords, including a supervisor password. The hackers deleted one NetWare account and some files but did no irreparable harm.

This problem is fixed in NetWare 4.X because the workstation shell includes integrated support of AppleTalk protocols that will pass network addresses to administrators.

NetWare Virus Control

A virus in the public area of a NetWare server will infect many workstations in minutes. By the time virus scanners such as Cheyenne's Inoculan or Intel's LANProtect notice, it is too late.[3]

Since viruses attach themselves to executable files, it is best to use NetWare's trustee rights to give users read-only access. In this way an infected PC cannot infect a server-based program file.

The back door

If you log in from an infected PC as the supervisor, you can unwittingly infect every server-based file, regardless of their user's rights. The key is to avoid using the supervisor ID as much as possible. Instead, manage user connections from the console. Create a print queue operator for someone other than the supervisor.

Another technique is to create an ID with workgroup manager status. The workgroup manager can create, delete, and modify accounts without access to the entire server.

Designate a user as workgroup manager from SYSCON using Supervisor Options. In User Information, pick the new workgroup manager's ID. Use the Managed Users and Groups option to make all other users and groups fall under the workgroup manager. Any new accounts created by the manager will give the manager the same rights over them.[4]

Virus-scanning NLMs

Third-party vendors such as Cheyenne (InocuLAN), Intel PC Enhancement (LANProtect), Central Point (Anti-Virus for NetWare), and McAfee (NetShield) market NetWare Loadable Modules (NLMs) that scan for viruses when programs are launched (see Fig. 8.3). Since these NLMs run in real time, they delay transfer of the file to the workstation until the scan is complete. This means a delay in getting the file out plus an added burden on the server. Consequently, such NLMs can take a toll on server performance. The more files to be opened, the worse the performance. NetWare SYSCON opens six files. Windows, however, opens 165.

Traffic load affects scan speed because both activities contend for CPU cycles. Heavily loaded CPUs will bog down still further with

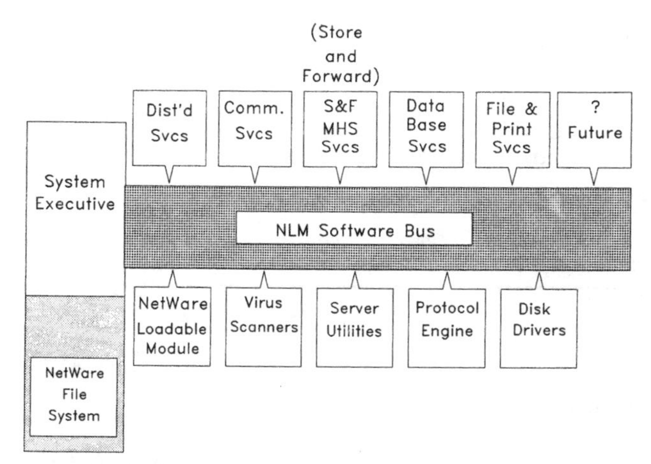

Figure 8.3 NetWare Loadable Modules.

the added NLM burden. On the other hand, lightly loaded machines may not even notice.

The key is to test. Try opening a Windows application from a workstation with the NLM running and without to see if there is a difference. To improve performance, most NLMs will allow selective scans of volumes and .EXE and .COM files.

The NLM itself should not pose a security risk. If not well behaved, an NLM can interfere with other NLMs, conflict with other server tasks, or even crash a server.

There are no standards for virus-scanning NLMs, so expect a wide variance in capabilities, quality, and performance effects. A full discussion of virus-scanning techniques is found in Chapter 2.

Workstation-based versus server-based scanning

As a perimeter defense, most of these NLMs also come with virus-scanning TSR programs for workstations. This is the best of both worlds. Workstation scanning has its advantages and disadvantages as shown in Table 8.3.

Server-based scanning

As Fig. 8.4 shows, server-based scanning can be run as a one-time batch program or in real time. They protect only files on the server. Users are not involved in the scan process. They are useful for the following functions or situations:

- Continuous online scanning
- Many users or a changing user community
- When many foreign machines such as notebooks are routinely attached
- When outside disks are often inserted into workstations
- When LAN application software changes often

TABLE 8.3 Advantages and Disadvantages of Workstation Scanning

Advantages	Disadvantages
Low cost	Users must be taught to use the scanners
Versatile and portable: Use in laptops, notebooks, PCs	Workstation scanners must be updated periodically
Can find and sanitize infected machines	Indirect control over user scan implementation

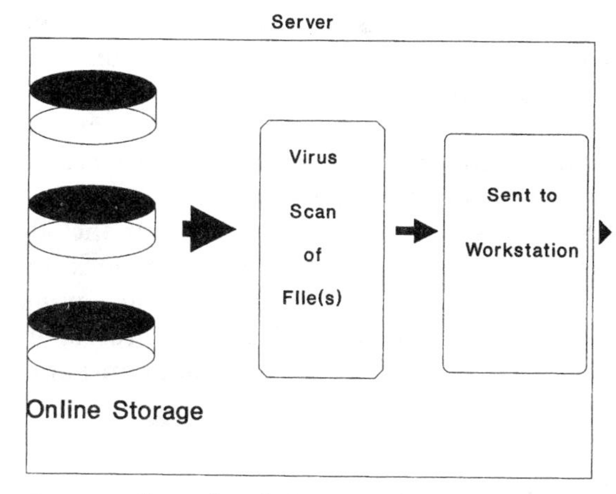

Figure 8.4 Server-based virus scanning.

Additional Security Features in NetWare 4.X

The release of NetWare 4.X improved NetWare's already good security even more. If you are a current NetWare 3.X user, it is important to understand these features because Novell plans to migrate many of them from 4.X to 3.X. This has already happened with respect to System Fault Tolerance III (see Fig. 8.5), which backs up servers in real time.

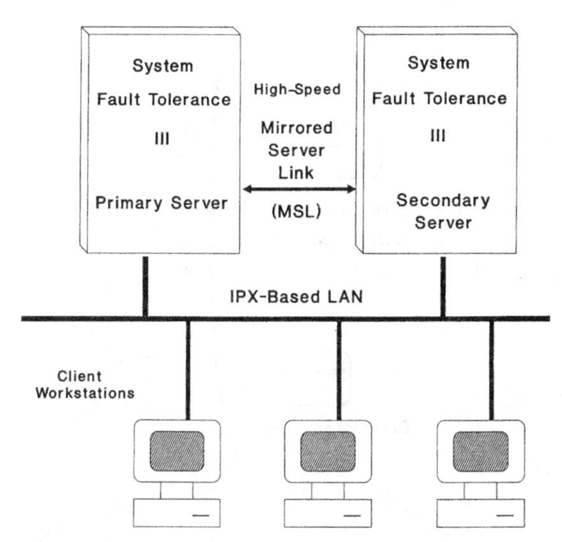

Figure 8.5 System Fault Tolerance III.

In the 3.X version of NetWare, it was necessary to administer access to each server (through the bindery) individually. This was difficult and often led to excessively liberal rights. Additions such as Novell's NetWare Naming Service and Banyan's Enterprise Network Services consolidated the management of binderies across multiple servers, but did not materially add to NetWare's security.

The new NetWare 4.X adds NetWare Directory Services, including extra access controls that can surround the entire network. As an enterprise NOS, the user logs in once, and uses Rivest–Shamir–Adelman (RSA) public- and private-key protection to keep unauthorized users out. NetWare 4.X also adds an audit tool that we will discuss below.

NetWare Directory Services (NDS)

The key to NetWare 4.X—literally—is NDS, as illustrated in Fig. 8.6. NDS is a database containing information about all network resources: users, printers, servers, print queues—everything.

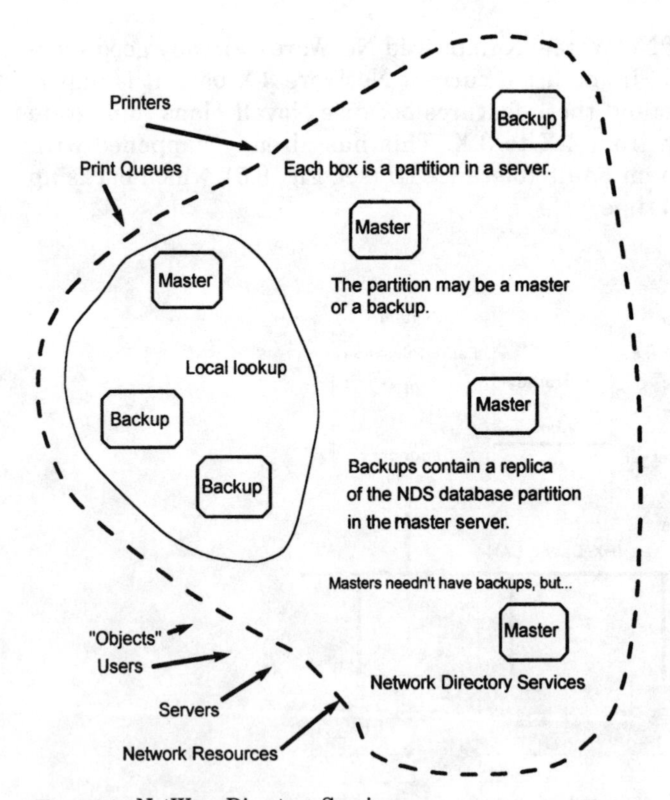

Figure 8.6 NetWare Directory Services.

Each resource is called an *object*. Objects are stored in the directory database. In fact, the database is divided into sections called *partitions*. The partitions improve performance because they are intended to be near the objects they manage. In this way, NDS does not have to traverse a wide area network to find an object, and the database in which to look is smaller.

The master copy of a partition is kept on one server, while other servers act as backups. This means that there must be at least one more server to back up the partition. Should the master server fail, the backup will recover. If there is no backup, then that section of NDS data—all the objects within it—will be inaccessible.

The virtue of this system is that the network manager defines an object and all its attributes only once. There is a single account per user and a single set of access rights for each account. Included within this definition are all the rights discussed earlier for *users*. NetWare 4.X also defines what actions a *program* can take.

NDS adds ten more access rights for objects plus the five rights we have seen in the NetWare 3.X bindery. While a useful extension, a larger scale—the entire enterprise network—must now be considered from a security standpoint when assigning these expanded rights. The net effect is that in NDS there are now file or directory rights and object rights that include bindery rights plus extensions. Not too different from the parties of SNMP, objects can have a unique property called an access control list that specifies how objects can access each other.

Auditing

As we said earlier, auditing was not part of earlier NetWare versions. Auditing is the metric by which managers measure the proper use of the system consistent with their organization's mission.

Specific operations may be covered by the audit. A group of network objects called a *container* is one auditable population. The other is server hard-disk volumes. The manager sets one password for each and gives control to the auditor.

The auditor then changes the passwords so tampering cannot occur and sets up the activities to be watched. The passwords are required to get to the acquired data and cannot be changed. Like any other object, auditors must have access rights. No rights means no access.

If the file into which audit data gets filled, auditing can be turned off or the volume dismounted. If dismounted, the data are lost.

Auditing will surely collect an incredible amount of data. As yet, no tools exist to reduce it to a useful form. A second type of currently nonexistent data reduction tool would be useful for performance analysis and for highlighting improper activities.

If the manager gets run over by a truck, audit data become inaccessible unless he or she has foresightedly left them in a sealed envelope and someone knows where to look. The same passwords are needed to remount a volume dismounted due to overflow.

File compression and antivirus scanning conflicts

NetWare 4.X compresses files that have not been opened for a number of manager-defined days. This creates problems for antivirus NLMs and CRC-based integrity checkers. For one thing, the file must be decompressed first before scanning, adding to the time needed to scan the file, not to mention processing horsepower.

Another snag is that decompressing the file resets the compression utility clock. This means that the virus scanner must run less often than the automatic compression algorithm or no files will be compressed. Scanners will reduce usable disk capacity unless they cause the file to be recompressed after scanning.

File migration

When the server's free disk space gets too low, NetWare 4.X migrates inactive files to a tape or write once–read many (WORM) drive until space is once again free for them on the main drive. So as not to scare people to death, the file names and sizes remain visible, but the data are really nearline (neither completely online nor totally offline). If a migrated file is called, it is pulled from near line to online.

From a security standpoint, a nearline machine failure means no access until the device is fixed. It is desirable to exclude certain files from nearline storage if they are considered crucial. Of course, this presumes that backups are performed on the server drive(s). Neither nearline nor offline storage is *backup* storage.

Storage management services

Storage Management Services (SMS) was a Novell product long before NetWare 4.X, dating back to 1991. Novell hoped it would become an industry standard, and indeed it has gained the support of Hewlett-Packard and DEC as a generic method of backing up data across a network regardless of vendors, NOSs, media, or platforms used.

SMS is a series of software modules that permits a server to back up and restore not only itself, but other servers, workstations, and especially NetWare 4.X's network directory services. This is the special virtue of SMS, as no other backup system can currently preserve NDS data. SMS can back up server-located files in DOS, file transfer and access management (FTAM), network file system, and server-based OS/2 plus OS/2 and DOS data from workstations. It can also do full, incremental, or differential backups.

Unfortunately, SMS needs a lot of server memory and supports only tape drives, not tape changers. However, its implementation is essential if NDS data must be backed up (e.g., there is no replicated server in which to store a copy of the NDS partition).

NetWare Core Protocol Packet Signature

A student at the University of Leiden in the Netherlands wrote a spoofing program for NetWare 3.11 called HACK.EXE in late 1992 that monitors packets flowing between the workstation and the server. By copying the packet's details, it was possible for the program to pretend to be the legitimate workstation.

The server could not tell the difference, and so the spoofer acquired whatever rights that were possessed by the workstation. This even included the network supervisor. (Here is another reason to log in as supervisor only from the server console and avoid using RCONSOLE whenever possible).

To plug this leak, Novell introduced a fix in November, 1992 called NCP Packet Signature (Fig. 8.7). Using RSA public-key technology, Packet Signature attaches an eight-byte identifier to the end of an IPX packet as an authenticator. NCP Packet Signature consists of an NLM on the server, modified client shell software, and user login utilities. A NetWare 3.11 utility, WSUPDATE, lets managers install client software via the network.

NCP Packet Signature became available first for NetWare 3.11 as an NLM (and is built into NetWare 3.12), then was used for NetWare 2.2 and is built into NetWare 4.X. It is also in NetWare for Unix. The fix is available through NetWare resellers, on NetWire (via CompuServe), and NetWare Express or by calling Novell at (800) 638-9273.

Signatures change by *transaction,* not by *session,* so copying the signature does not permit spoofing. Signatures can be used by some or all workstations in a mixed environment and by the server,

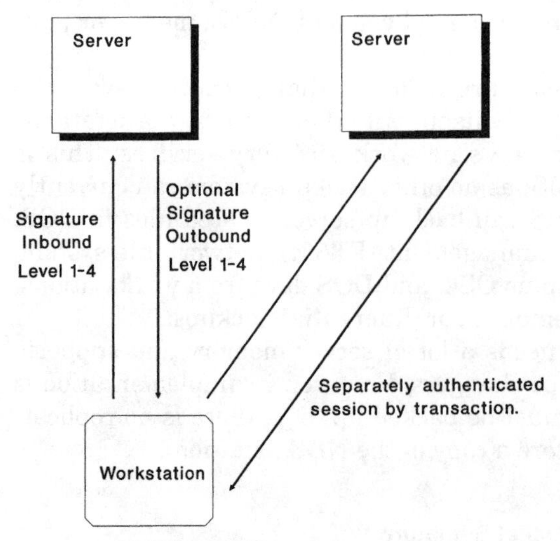

Figure 8.7 NCP Packet Signature.

too, if two-way authentication is desired. As an aside, you will enhance security by limiting the number of concurrent sessions allowed a user to a small number, perhaps two. NetWare permits up to eight.

Multiserver networks can each have different packet signature levels. Authentication is by server, not network. Higher-signature-level security is possible by server, so that machines with sensitive data can get more protection than those containing less sensitive information. In all, there are four signature levels.

Servers can be configured to work with both signature and nonsignature workstations, allowing partial migration if desired. Incorrectly signed packets cause an alert message to be sent to the client, the server console, and to the error log. The client–server connection is not broken.

NCP Packet Signature has a few negatives (also see Table 8.4):

- Performance degradation in small to midsize networks is small. Larger systems of 250 machines or more will see about a 10 percent throughput degradation.
- Packet Signature only applies to NCP sessions using IPX. Protocols such as AppleTalk File Protocol, TCP/IP, and Xerox XNS are *not* protected.
- *Data* are not encrypted. Novell is expected to announce an NLM to encrypt IPX packets this year. Again, non-IPX protocols are not covered.

TABLE 8.4 When Should One Use Packet Signature?

Do not use if:	Use if:
There are only executable programs on the server.	Workstation user(s) are not very trustworthy.
You know and trust all workstation users on your network.	Physical access to the LAN cable is easy, giving a spoofer an entry point.
Data on the server are not sensitive.	You have publicly accessible workstations or workstations are left unattended.
Loss or corruption will not affect you.	

Security of NLMs

It has long been known that application NLMs can corrupt data because they can intrude on memory reserved for the server. Improperly written, the NLM can crash the server.

NetWare 3.1X and all NLMs for it run at privilege level ring 0, the least protected level. A misbehaving NLM can write over memory not its own and cause a crash and possible data loss. In NetWare 4.X, this problem is fixed. NLMs run in ring 3 (OS-protected domain), the most protected ring. NLMs venturing out of their domain are caught and terminated. While there is a small performance loss with this method, it is well worth it.

The Riverbend Group Breach

In March 1993, a Novell instructor with the Riverbend Group of McLean, Va. accidentally set a supervisor password during a training session. He was then faced with finding a way around it in order to regain system control.

His solution was ingenious. He first dismounted the SYS. volume at the server, which contains all the core NetWare operating information. He then renamed it something else and brought it back up. The system thought it was a new volume and asked for a new password. The instructor then entered a password and regained full supervisory access to the network and server. The entire process took under a minute.

Lessons to be learned

This would not have been possible had:

- The server been locked up

- A console password been in use
- The keyboard been locked
- The "attacker" been less knowledgeable

Bypassing Novell Login Scripts

Most Novell networks and some other LANs use login scripts to automate the login process. Just before login, at what is called the *attach point,* Novell security can be breached. This is done by typing IPX, NET, and ATTACH. If entered at the right time, many Novell scripts and security programs will not run. However, the user will attach to the server. There is no easy defense for this oddity, save for "security through obscurity."

Novell Summary

The user community gives NetWare high marks for security (see Table 8.5). Novell has shown that it addresses security holes quickly and aggressively, as they did with the Leiden breach. Novell also has plans to certify NetWare under the National Computer Security Center Trusted System program at the C2 level, considered industrial-grade security. Novell has further shown its security commitment by making it such an integral part of NetWare 4.X. However, these security features are useless if they are not implemented.

For a listing of Novell security utilities, see the end of the chapter.

Banyan Virtual Network Services (VINES)

VINES was intended from its outset in 1984 to be an enterprise-wide NOS. Until NetWare 4.X, it had virtually no competition at the enterprise level.

Rather than describe all the security features of VINES, many of which parallel those of NetWare, we will focus on the unique strengths and a few weaknesses of VINES.

Authentication services

Like NetWare 4.X's Network Directory Services, VINES has its own set of directory services called StreetTalk (see Fig. 8.8). StreetTalk exists in VINES servers. When a server joins the network, its StreetTalk tells everyone else about the groups and

TABLE 8.5 Novell Network Security Recommendations

- ✓ Use only the most current versions of system, client, and patch software.
- ✓ Check periodically for viruses.
- ✓ Use the Security utility to detect vulnerable server access points.
- ✓ Lock servers in a secure room.
- ✓ Issue the SECURE CONSOLE command from the NetWare console. Then NLMs can only be loaded from SYS:SYSTEM.
- ✓ Activate the lock file server console from the monitor main menu when the NetWare console is not being used.
- ✓ Always use a password different from the supervisor password for RCONSOLE.
- ✓ Limit the number of users possessing supervisory rights. Use a workgroup manager instead.
- ✓ Avoid logging in as supervisor whenever possible.
- ✓ Avoid assigning security equivalences that give one user the same access rights as another user. If you must, revoke it as soon as possible.
- ✓ Use access control features to limit not only users' but virus' access to data and applications (directory level).
- ✓ Enable intruder detection and lockout.
- ✓ Ask users to log out whenever they leave their workstations, or install a product like NetOFF that logs them out after inactivity.
- ✓ Lock up, disable, or otherwise secure unattended workstations.
- ✓ Require passwords for all accounts of at least five (Novell) characters. At least six mixed characters are recommended.
- ✓ Make users change their passwords at least every 90 days.
- ✓ Require unique passwords, as we have discussed.
- ✓ Limit the number of grace logins (login attempts).
- ✓ Limit the number of concurrent connections in NetWare 2.X or 3.X. NetWare's maximum is eight concurrent workstation to server connections.
- ✓ Enforce network login time restrictions and station restrictions.
- ✓ Train users and administrators how to use NetWare security features.

resources within it. The other servers tell it their groups and resources. But detailed authentication information for each user and resource exists only in the server who owns that user or resource.

Should the server fail, there is no replication of these authenticators at all, so these resources and users would not be available, nor would users have access. Recall that in NetWare 4.X, if there is

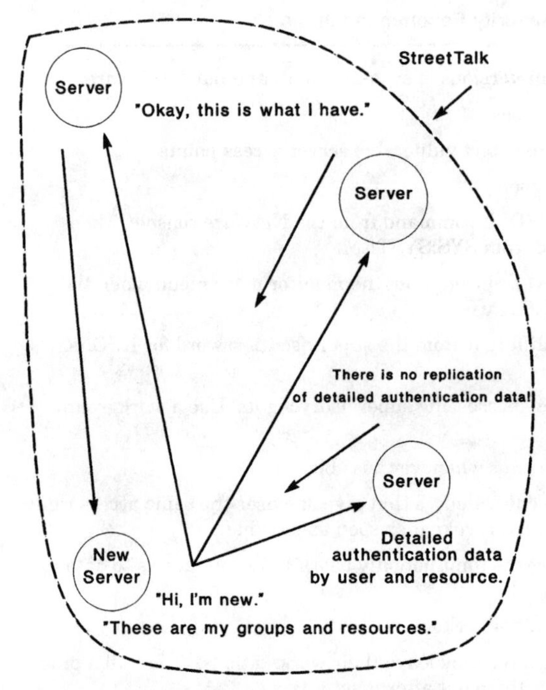

Figure 8.8 VINES StreetTalk directory services.

more than one server, it (or they) will back up the NDS partition as hot spares. However, there is no way to back up the NDS *database* except to Novell's SMS backup system.

Banyan planned to add StreetTalk fault tolerance by the end of 1994. The plan was to allow up to two shadow servers to replicate StreetTalk user access rights only, rather than Novell's approach, which is full server replication. Banyan's position is that full replication could lead to security holes by moving and storing more information than is actually needed. Suggestion: Ask your vendor about Banyan's shadow feature before you buy.

VINES user controls

VINES logon authentication defaults are not sufficiently restrictive. Users should not therefore accept the defaults, but rather make them more restrictive. Banyan advises this themselves.

DOS, Windows, and OS/2 clients' passwords are encrypted, always. Macintosh passwords are encrypted in StreetTalk, but their passwords are not encrypted if AppleTalk File Protocol (AFP) is used. AFP is the native network support of the Macintosh operating system.

VINES offers the ability to encrypt all traffic via internal software if desired. They are the only NOS vendor to presently offer this capability.

Disk space usage

It is standard practice to allocate maximum disk space by user in a large network. If not done, a "disk hog" can take up disk space required by the system itself, slowing service to users. NetWare offers this feature, but presently VINES does not.

Rights

VINES rights, like NetWare rights, are sophisticated, flexible, and complicated. The antivirus principles that apply to NetWare rights apply equally to VINES. Similarly, the file and directory attributes are just as complex and require experience and planning to use them as a security-building tool.

Unlike NetWare, VINES has few utilities to help out in this area. This is because Banyan was until recently a semiclosed environment and because VINES has under 10 percent of the NOS market to Novell's 64 percent.

Interserver security

VINES excels in its ability to limit the type of access that can occur between networked servers. This feature is unique to VINES. The preferred implementation method is to secure each machine fully and then set the proper access level for each interconnected server.

Auditing

Like NetWare 4.X, VINES has auditing capability to detect logon attempts, security violations, and the like. Also like NetWare, there are no data reduction tools to make sense out of all the data it produces.

"Dongles"

VINES uses plug-in hardware keys called "dongles" or "birds" that enable use of its software in servers. Plugged into the first parallel port, VINES software interrogates it to see which services are allowed. The parallel port can still be used to attach a printer.

New services require another dongle to be cascaded behind the first. Once an upgrade is complete, the dongle(s) may be disconnected and returned to Banyan for credit.

These devices are obsolescent. They can fail or be lost, damaged, or stolen. They add unneeded complexity and create another vulnerability. Other vendors enforce licensing rights through their software. There is no reason for VINES to be different.

VINES's future

It is clear that StreetTalk needs fault tolerance. It is unacceptable to have resources and users offline until a server gets fixed. Banyan is working on StreetTalk's shadow backup, and it may well be available now.

Banyan had planned to submit VINES to the National Computer Security Center to gain a C2 government security rating. Banyan later put the project on hold pending enhancements that would affect the project. If C2 security is important to you, ask your vendor regarding its status; Banyan did say that they remained committed to gaining C2 security for VINES.

NetWare 4.X offers System Fault Tolerance III, which means online duplexed servers. Banyan offers no such capability at present. Banyan does offer RAID level I duplexed disks, but none higher as yet. All levels of RAID (1 to 5) can be accommodated in NetWare 3.X or 4.X.

The Banyan development environment had been relatively closed to third-party developers. This has been a matter of Banyan attitude and the fact that they have a seventh of the market compared to Novell.

The result for users is that the wealth of NLMs for security, virus detection, and everything else is largely missing in VINES. That is changing, but third-party support remains well behind NetWare.

Microsoft LAN Manager

LAN Manager is a holdover product. Based on OS/2, it is highly unlikely that the company will expend further effort on it than necessary to maintain present customers. Microsoft is pressing LAN Manager customers to upgrade to Windows NT Advanced Server.

The new LAN Manager is Windows NT [Advanced] Server (NTAS), although it still requires LAN Manager's presentation format. Based on Microsoft's Windows NT, [Advanced] Server is the latest in a new series of NOSs.

For the moment, though, about 10 percent of NOSs are using LAN Manager. In the same spirit as VINES, let us look at LAN Manager's security strengths and weaknesses.

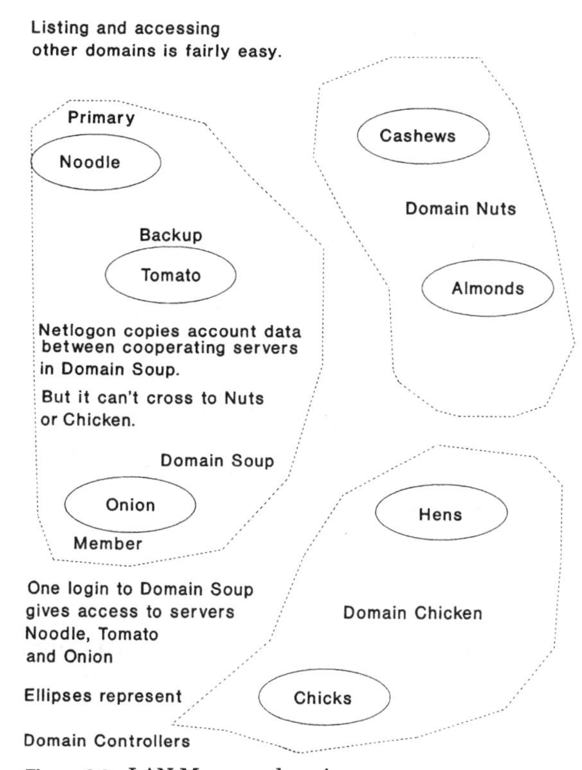

Figure 8.9 LAN Manager domains.

Domain-based architectures

LAN Manager is based on *domains* (Fig. 8.9), groups of servers where one login will suffice. Domains are an older NOS concept that was implemented in products like LAN Server (which is closely related to LAN Manager) and DEC PathWorks, whose origins LAN Manager shares.

A domain is not a global directory. One might think that domains implement the need-to-know principle by dividing the network into parts. In reality, it is easy to list resources in other domains, and get to them. The bottom line is that domains are cumbersome to work with and add little to security.

Management of the domains is easier if a program called *Netlogon* is running. Netlogon copies user account data between cooperating servers in a domain. If the primary domain controller fails, logons can be validated by backup controllers. Lower-level member controllers hold user account data but cannot validate logons.

The problem is that even Netlogon allows no logical connection between domains. This means that each domain must be managed individually. Consequently, management is easier if users are limited to just one domain.

Share-level security

The less sophisticated of two models that LAN Manager contains, share-level security defines a sharename for a resource, and then a password to get to the sharename. Access permissions define what can be done with the resource specified by the sharename. By using different passwords with different access permissions, several layers of access can be created for the same sharename. The result is that many users wind up having the same password (always a bad idea). While it might be all right for purely generic resources, using LAN Manager seems an expensive way to do it.

User-level security

This is the alternative to share level and has much finer resolution. Here, specific resources are allowed for each user.

Encryption

In order to encrypt passwords on MS-DOS client workstations, an optional service called Encrypt must be installed and running. Otherwise new passwords selected at the workstation will not be encrypted. This is rather appalling. There should be *no* choice here. *All* passwords should be encrypted.

If LAN administration is done remotely, away from the server, Microsoft recommends that it be done from a Windows or OS/2 workstation. The reason is that on other platforms (except as noted), passwords travel from workstation to server *unencrypted.*

LAN Manager comes with two default named accounts, ADMIN and GUEST. GUEST does not need a password and is intended for those simply using printers and such. Regardless, it should be protected by password anyway.

Securing resources

A program called *Netrun* accesses the resource needed to execute a program on the server. Although Netrun can be restricted, it might be possible to run an unsafe program from a safe one. Also, general applications can be extended using macros to do entirely new and uncontrolled functions beyond system controls.

While most server programs cannot be run using Netrun, OS/2 command line applications using STDOUT will work. When all is said and done, the opportunity for a breach is just a little too great.

Permissions

Execute permission allows clients to execute applications and nothing else in a server directory. However, it can only be compelled for clients using MS-DOS 5.0 or higher or OS/2. For pre-MS-DOS 5.0 versions, Windows, and Macintosh, the selected application will not be executable and the looser read permission must be used (read allows copying, so it is weaker).

Users get permissions for a resource from all groups to which they belong. If a user has read and write permission in one group and read and create in another, that user gets all three. However, individual permissions can be set to override group assignment. When many large groups of users are defined as in large servers, this becomes a difficult problem to manage.

If the standard file allocation table is used for disk partitions, new subdirectories have no permissions assigned to them. On the other hand, partitions using the OS/2 High Performance File System (HPFS) inherit the permissions of the parent directory. Microsoft recommends HPFS.

Audit

LAN Manager has a respectable audit function, but like VINES and NetWare 4.X, has no reduction tool to turn *data* into useful *information.*

Summary

LAN Manager has gained some good reviews and is implemented in some very large firms. It is nicely integrated with other Microsoft products: Windows for Workgroups, MS-DOS, and Windows; even LAN Manager for Santa Cruz Operation (SCO) systems. The SCO supports LAN Manager under Unix. If you have LAN Manager in your IS environment, make use of all the security tools it offers. As Microsoft draws support away from LAN Manager, users will be forced to consider Windows NTAS more seriously.

DOS

Various hacker tools have been written that can penetrate almost any DOS-based system. Oddly, the most often used methods of penetrating NetWare LANs are not NetWare specific. This circumstance is a variation of "Maginot Line thinking," where Novell built stout controls that hackers beat by simply taking a different approach.

Be alert to three programs that hackers can use:

- NETCRACK is a program that finds user passwords on NetWare LANs.
- THIEF and GETIT are two programs that capture NetWare passwords and IDs before they are encrypted.
- KEY COPY copies every key stroke entered by a user on their PC.

Unchanged passwords recorded as keystrokes by hackers gain them entry before NetWare controls are invoked.

The defense to these programs is a token password as would be generated by a system like SecurID. Then recorded keystrokes become useless.

Unix and OS/2

Some versions of Unix and OS/2 will return a superuser prompt if the input buffer is grossly overloaded by a very long message. Test your system to see if this can be done.

Unix is a fine operating system, but it was never designed for great security. Only one version of Unix, System 4.2, has earned an NSA B2 classification.

The problem with Unix is that it is mainly a peer-to-peer system. As such, access privileges are determined on a station-by-station basis. It is easy to get a copy of the file containing the account names and (encrypted) passwords. Once in a cracker's hands, they can break the passwords at leisure.

CERT suggests the following:

- Check your accounting, C2, and system logs. Unusual entries or activity signal trouble. Nothing unusual may also signal trouble if the cracker has tampered with the logs.
- Hidden files are sometimes created by hackers as places to store their ill-gotten gains. Odd, cryptic names are often a giveaway, such as dot-dot space or dot-dot-control-z (or some other letter).

- Run this command to find files that set the user ID to root or some other privileged account: find / -user root -perm -4000 -print.
- Trojan Horses in Unix take the form of Unix commands that actually contain something else. These might include popular TCP/IP commands such as *Telnet, ftp,* or *finger.* Unix commands such as *su* (substitute user) or *login* are popular for this purpose. Make sure the file lengths are the same as the lengths you recall as the real thing.
- Look at the path. Unusual or unfamiliar subdirectories, especially at the beginning, imply trouble.
- Check the *cron* and *at* control files. They should run only programs you recognize. Files referenced by *cron* or *at* should not be writable by everyone.
- Similarly, keep an eye on the */etc/passwd* file. Accounts without passwords, new accounts, and user IDs bear special watching.

Windows NT

Microsoft intends to certify Windows NT at the federal C2 level first and later at the higher B2 level. If you need this level of security, request the status of the certification effort from your vendor. Such certification can take years.

Apple System 7

System 7 file sharing lets any Macintosh computer be a file server. If a user turns on file sharing to do a transfer, turns on guest access, and later forgets to turn it off, his or her machine is open to everyone. Solutions to this problem include:

- Remove the sharing setup control panel from each Macintosh. While secure, this solution feels like throwing out the baby with the bathwater.
- Better still, teach your users to set up a share folder instead with guest access. AG Group's [(510) 937-7900] Nok Nok ($50) software advises you whenever anyone connects to your machine to do personal file sharing. Another product, ultraSecure ($239) from usrEZ [(714) 756-5140] does the same thing but has more features.

- Inventory your system with Network Security Guard ($259) from MR Mac Software. It will find out which servers are running what software; which have guest access and which ones have easily-broken passwords.

Later versions of System 7, specifically System 7.5 and System 7 Pro have more security options. One is the Apple Open Collaboration Environment (AOCE). AOCE includes a keychain holding several user IDs and passwords. All are encrypted until unlocked by one password from the user.

The problem is that once unlocked, individual passwords are sent around the network in plaintext by many network servers. Tapping into these passwords is a snap. The main benefit of AOCE is that it keeps people from writing passwords down or placing them in accessible documents such as Preferences.

AppleShare

There is not much security in AppleShare. Repeated attempts at entry will not be cut off as with NetWare's Intruder Detection and Lockout. Nor does AppleShare allow a manager to specify who can connect to what resource at what location during a given time of day, or during a given day for that matter.

One partial solution is to use *Nok Nok A/S* to add logging and restrict the time that active, idle, and guest users can connect to the server. Nok Nok A/S identifies guest users by their machine names, and will alert you when someone logs on to AppleShare (if you are near the server and can see or hear it).

For better protection, switch to AppleTalk Filing Protocol as a file-sharing system. Its cost is greater, but its security is on a par with NetWare for Macintosh, VINES Option for Macintosh, and DEC PathWorks. The administrative workload is a significant hidden cost that goes along with a more sophisticated NOS.

PowerBooks and Macs can be stolen like any other machine. For a list of Macintosh desktop security software, see the end of the chapter.

General OS Security Notes

Electronic mail

It is easy to forge electronic mail messages. This is why products such as Lotus Notes offer digital signing or encryption. If a message seems out of character, it could well be a fake.

Time of check to time of use (TOCTTOU)

Between the time an operating system checks security and the time it begins a process, security information may be stored in memory and therefore subject to modification by an attacker. This is called a TOCTTOU attack.

Superzapping

Sometimes a computer will freeze or lock up, and a *superzapping* routine will be used to thaw it out. These routines can sometimes bypass normal security controls.

Browsing

Often, memory given to a new user will not have been erased by the last user or by a process designed to wipe the memory clear before the next user logs in. The new user can browse the first user's memory for useful information they should not have.

This same principle applies to erased files. The actual data are not erased, only the name. Tools like Norton Disk Doctor can read and recover this data. Some operating systems are designed to randomize erased file and memory space. Utilities can also be bought that do this.

A successful security breach occurred when an adversary read "scratch" (supposedly empty) tapes that contained useful seismic data. This went on for months before it was caught.

The salami attack

A few cents here...a few cents there...no one will ever notice. That is the essence of a salami attack. By stealing the roundoff of an interest payment, for instance, a thief can steal thousands or millions over several hundred thousand customers. This is more of an applications program than an OS breach, but it is worth mentioning.

Review

- We underestimate the value of information by mistaking it for the value of the system that contains it.
- For all machines, disabling floppy drives adds to security.
- Review your insurance coverage so you know what is and what is not covered, and how to present a credible claim.

- The security plan is backed up by the disaster recovery plan.
- Novell NetWare has very good security, *if it is used* (as with other NOSs).
- In all NOSs, restricting rights and permissions adds security and can prevent the spread of a virus.
- Server-based virus scanning and workstation-based scanning each have their merits. There is nothing wrong with using both.
- Novell's security recommendations apply nearly equally to all NOSs.
- Banyan VINES is also a secure NOS, with some notable weak spots in authentication backup and disk space allocation.
- Microsoft LAN Manager has a number of weaknesses, several having to do with unencrypted passwords. It is unlikely that these will be fixed as Microsoft has moved toward its next generation NOS, Windows NT Advanced Server.
- Common to all NOSs, the audit process generates reams of data but very little usable information. Data reduction tools do not as yet exist.
- Several DOS-based programs exist to copy passwords. Use a token password system to get around these programs.
- NetWare security and login scripts can be bypassed by a certain sequence of commands prior to the server attach point.
- Unerased media, be it memory, disk space, or tape, can create an inadvertent security breach. This is called *browsing*.
- Some operating systems can be breached using TOCTTOU or superzapping.

Appendix. Netware and Macintosh Security

NetWare security software utilities

Cheyenne Software, Inc.
Roslyn, N.Y.
(800) CHEY-INC
NetBack (also called Cheyenne Utilities)

Frye Computer Systems
(800) 234-FRYE
Frye Utilities for NetWare

Intrusion Detection, Inc.
New York, NY
(212) 360-6104
Kane Security Analyst, analyzes NetWare weak spots

LAN Support Group
Houston, Tex.
(800) 749-8439
NetSqueeze + Encryption offers data compression and optional encryption.

ON Technology
Cambridge, Mass.
(617) 374-1400
AudiTrak, also a NetWare auditor

Software Inc.
Portland, Or.
(503) 294-7025
SmartPass (NLM), a password manager, available from this supplier and others.

Brightwork Development
(800) 866-6585
Brightworks/McAfee Tool Suite

Symantec
(408) 253-9600
Norton Network Administration Suite

Macintosh desktop security software

These products are made by Kent Marsh:

FolderboltPro is for single-user Macintoshes. It locks individual folders via password such that they can be set for read or write only, or no access at all.

NightWatch II is activated by events, such as shutdown or sleep. It then requires a password to do anything new.

CryptoMatic encrypts files using ANSI DES. Decryption calls for a password.

Magna *Empower Remote* and usrEZ's *ultraSecure* include all of the above and then some. All the above offer both DES and non-DES encryption for faster speed. Note that this is your only real protection for critical information; breaking an encryption algorithm is harder than unlocking a disk.

Network versions of the above packages are available.

Notes

1 The Network World BBS telephone number is (508) 620-1160 from 300 to 2,400 bits/s. Set 8 bits, no parity, and one stop bit (8N1). For speeds up to 9,600 bits/s, call (508) 620-1178 (also 8N1).

2 Leprechaun Software is at (404) 971-8900; fax (404) 971-8828.

3 Edward Liebing, "How to Use NetWare's File Attributes to Provide Security," *LAN Times*, December 7, 1992.

4 "Network Supervisor or Typhoid Mary?," Corporate Computing, May, 1993.

Chapter

9

Sources of Leaks in LANs and WANs

Objectives

At the end of this chapter you will be able to:

- List the most common threats to LANs and WANs
- Discuss forms of interception in wired and wireless systems
- Itemize the finer points of dial-in security procedures
- Describe techniques for improving packet switching network security
- Explain how leased lines can be compromised
- Show how bridges dilute security
- Indicate how routers can add security
- Identify which router protocols can be risky
- Show how gateways can be used to breach a network
- Outline secure backup procedures
- Understand how removable media can effectively connect two isolated networks
- List ways to anchor computer equipment physically
- Place the entire computer security topic where it belongs: in the people perspective

Introduction

Between the years 1988 and 1991, the number of reported security breaches on the Internet, the world's largest data network, increased from 6 to 405, a 67-fold increase in four years.

Captain Midnight made his debut on the HBO channel of cable TV when he invaded their satellite TV uplink in 1986. Home Box Office subscribers across the country found themselves paying to hear his harebrained political opinions.

The FBI caught 400 illicit wiretappers last year. There was no word on how many there *really* were.

Financial institutions are special targets. When asked why he robbed banks, Willie Sutton said, "That's where the money is."

Threats to networks come in five general forms:

- *Interception* of information.
- *Alteration* as with replaying data packets, masquerading as someone else, intercepting signoffs, or sending false messages.
- *Jamming,* primarily a problem with radio-based systems.
- *Denial of service* as could be caused by a virus or sabotage.
- *Loss of network integrity,* as with a major outage due to a cable cut or the Signaling System 7 failure experienced by AT&T that caused a massive voice network outage.

Interception: Cable Taps and LAN and WAN Monitors

Physical taps may be detectable using sensors of various kinds, but inductive pickups with no physical connection will be harder to catch. LAN and WAN monitors should be locked up. Unfortunately, any PC can become a network analyzer with the proper software. One version called *IPX Permissive* is often found on hacker bulletin boards. It can read Novell LAN packets and reprogram a workstation to any desired address. Some inventory programs can detect the presence of programs like IPX Permissive.

An analyzer may not even be needed. Some LAN network interface cards (NICs) and workstations allow their addresses to be modified so that they will accept every frame they see, not just the ones destined to the NIC's address.

Radio-Frequency Systems

Microwave links, satellite uplinks and downlinks, cellular phones, and wireless LANs are all vulnerable to interception. Frequencies

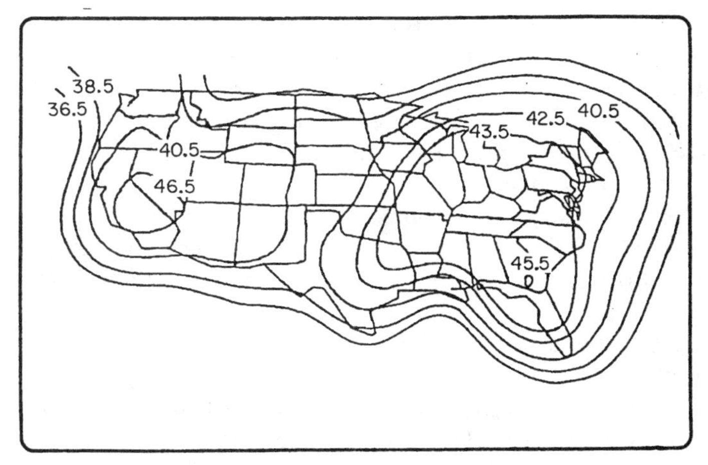

Figure 9.1 A satellite footprint indicating signal strength by area.

are published, data framing formats are standard and public, and equipment is inexpensive and readily available. As a stratagem, forget security through obscurity.

The National Security Agency (NSA) reportedly has listening stations near major carrier earth stations. The former Soviet embassy has line-of-sight visibility to all of Washington, D.C., and a forest of antennas on its roof.

As shown in Fig. 9.1, satellite downlinks have a "footprint" that illuminates a large portion of the earth. Receiving these signals is easy, and today the land is dotted with mushroomlike antennas doing just that.

Captain Midnight appeared uninvited on HBO by overriding HBO's satellite uplink signal. Premium cable TV downlinks from satellites are scrambled now to prevent theft of service.

More and more companies are using very small aperture terminal (VSAT) satellite transmitters and receivers, especially in the retail trade. The security of these links is vital to them.

Satellite attackers steal service by sending a very narrow-band signal up to the satellite that is then repeated through the satellite's *transponder* on the downlink. While not disturbing the main channel, the attackers are able to create a secondary channel link. The illicit link also robs the main signal of a small amount of transponder energy.

Satellites in geosynchronous orbit are threatened by more than 300 pieces of space junk *visible enough to see on radar.* In fact, satellites have had to be moved to avoid potential collisions.

Cellular phones are unencrypted and open, as Prince Charles learned to his embarrassment. While analog receivers were supposed to go off the market in early 1994, it remains easy for a technician to modify a receiver to receive cellular transmissions. Even tuning a UHF TV receiver to the high end of the band can sometimes catch cellular transmissions.

Cellular fax and the new Personal Communications Systems are just as vulnerable. (AT&T now offers Surity Transportable Cellular Telephone 9300 using Clipper-based encryption. The price is $1,695.)

Wireless LANs have varying degrees of vulnerability. Fixed-frequency systems such as Motorola's Altair are stationary targets. Spread-spectrum systems are harder to monitor because the frequency changes constantly. However, a receiver bought from the same vendor will work, and will be passive or active as the hacker decides. All wireless LAN vendors offer optional encryption. Buy it.

Dial-In Security

Here are a few more tricks to enhance dial security.

Modems that answer calls should delay presenting carrier tones for at least ten seconds. In this way, attack dialers will see no carrier and move on.

Do not use a banner for sign-on that identifies your organization or where you are located. Just show a message that advises the system is for authorized users only and that improper usage will be noted and intruders prosecuted to the fullest extent of the law.

Glaring

Test your dial-back system to see if it recognizes dial tone. When a caller dials in and enters the right authenticator, *the dial-in line should hang up*. A moment later, the dial-back machine should go off-hook, *recognize the dial tone,* and call the user back.

Some dialers mindlessly dial back *whether they see dial tone or not.* If the cracker stays on the line, dial tones will be ignored and the dial-back system will *think* it has reached the right caller. Crackers look for this kind of logical fault: they call it "glaring" (as in glaring unblinkingly at the port protector).

Encrypt important files. Some experts feel it is better to just encrypt everything. WAN analyzers can do everything LAN analyzers can do: encryption is your only protection against such attacks.

Disconnect unneeded connections to the outside world. You should know exactly how many modems are in your own as well as your users' hands. A *Network World* BBS program called LOGINCHK will detect a modem in a user's workstation when they log into the LAN.

Limit where a dial-in user can go once into your system. Make functions they do not need inaccessible. Obeying the need-to-know principle, do not give users *carte blanche*. Some modems used for dial-out can automatically answer as well. Automatic answering should be disabled if not needed.

Modem signals at 2,400 bits/s and below can be recorded. At the beginning of a transmission, these signals may well include an embedded password. Then they need only be played back by a masquerader to gain access. Token passwords are the only defense.

Packet Switches

Many of the successful attacks on computer systems were through dial-up modems and packet switches. The infamous "Dalton Gang" from the Dalton School in New York's Manhattan broke into two networks, one in the United States and another in Canada in 1980. The leader of the gang was all of 13 years old, and in the eighth grade.

In Illinois, the Argonne National Laboratory was penetrated via a packet network. Attackers used default passwords on installation tapes that had not been removed.

You will recall the anonymous firm that had addresses in its virtual private network installed by the vendor for maintenance and test purposes. The customer was not informed. Ultimately, a security audit revealed their presence.

In an effort to add to the security of their packet networks, BT North America and Infonet Services Corporation have added security offerings to their services. BT's Security Products Portfolio includes the following:

- User authentication is accomplished by PIN number and dynamic password.
- Data transmitted by BT's X.25 packet-switching services is encrypted internally by BT (see Fig. 9.2).
- Users receive daily, weekly, or monthly reports that track security violations and analyze traffic patterns.
- BT offers consulting services to assess users' voice and data networks.

The Portfolio works with any application running on BT's X.25 network, including X.25, TCP/IP, and frame relay. The cost is about $6 monthly per user.

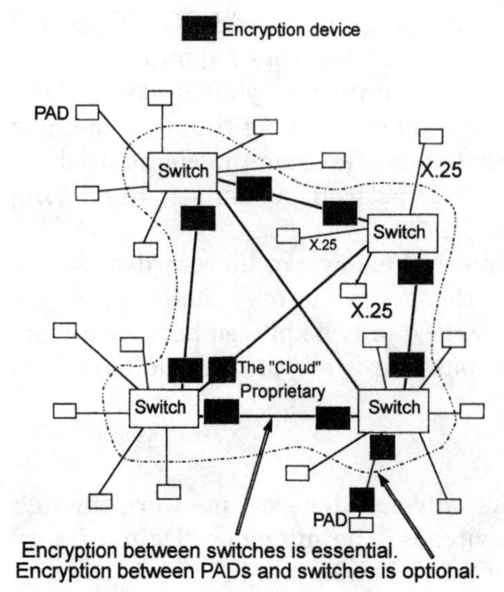

Figure 9.2 A packet-switching network. (A PAD is a packet assembler and disassembler.)

Another X.25 vendor, Infonet, offers a similar security service via its InfoLAN TCP/IP product. AT&T and Sprint currently offer only data encryption security service. CompuServe offers encrypted frame relay, a relative of ITU-TSS X.25.

These features are attractive to users because they avoid forcing users to buy equipment or divert staff energies to these security functions.

Leased Lines

Because they are static, leased lines are among the easiest to tap. Almost invariably, they are marked nicely and neatly by the telephone company. Find a circuit number beginning with FDDC or FDDN, and all you need is a modem and a recording device to steal information.

Demarcation points should be in a locked room, telco 66 blocks and RJ-21X junction blocks likewise. Encode circuit labels so that people cannot tell what they are. With their myriad of wires going everywhere, distribution frames are the perfect place to tap data. Lock them up.

T-1 or T-3 circuits have channel service units (CSUs) with convenient plug-in ports (see Fig. 9.3) for monitors. CSUs carry many channels of data, and their data formats are easily dissected. CSUs should be locked up along with any T-1 or T-3 analyzers in house.

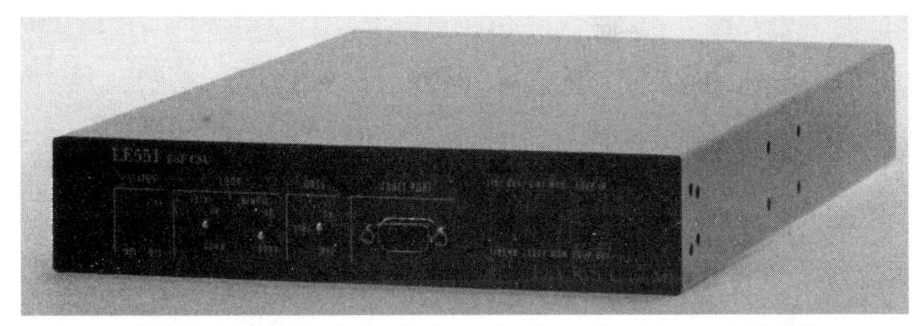

Figure 9.3 Many channel service units have front panel test ports (right) that provide easy access. (*Courtesy Larsecom Corp.*)

Network integrity is a concern. Until the 1984 divestiture of AT&T, telephone lines were routed according to government guidelines on integrity, in part due to survivability concerns in case of war. Today, outages are tracked by the Federal Communications Commission (FCC) and the Exchange Carriers Association. Outages of 30,000 lines or more must be reported to the FCC.

Bridges, Routers, and Gateways

All these machines should be locked up (see Fig. 9.4).

Bridges

Bridges can filter addresses that pass from one side to the other, but they cannot see what the attached information is. It could be a virus. If there is a maintenance port, it should be protected by a port protection device of the type described in Chapter 4. Even a simple asynchronous port meant for use directly with a VT-style terminal should be disabled when not in use, or it should be protected by password.

Bridges decrease system availability by passing along network problems from one segment to another. A broadcast storm on one segment will instantly cripple another. Routers, being more aware of protocol, can detect and mitigate a storm, and so isolate the problem.

Bridges can unlock a secure network if the bridge links it to an unsecured net. For instance, a caller dials in to a LAN and breaks the fairly simple password. Any bridges connected to the LAN will allow the intruder to hop from segment to segment without restriction unless address filters are installed on the bridge to disallow source and/or destination addresses from passing through part or

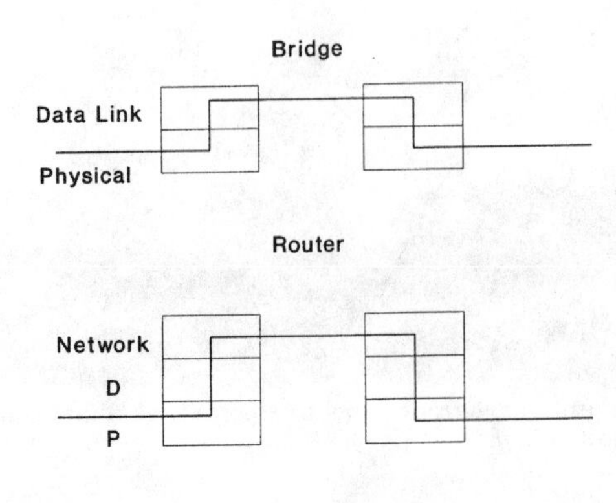

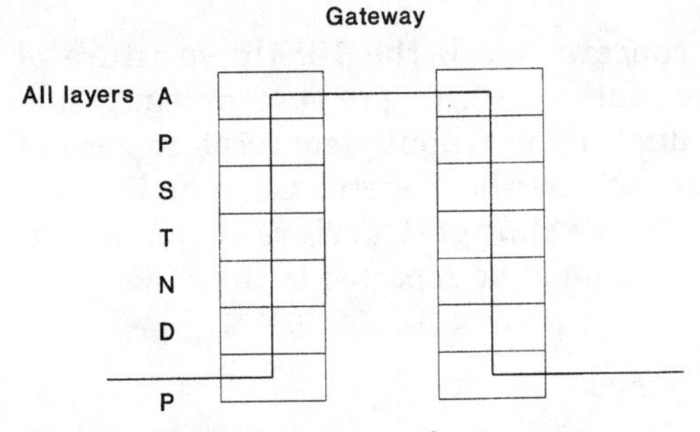

Figure 9.4 Bridges, routers, and gateways.

all of the network. In effect the intruder has bypassed whatever security barriers are in place on the adjacent LANs with respect to dial-in security. Unless neighboring LANs are encrypted or authenticated or bridge filters are enabled, the intruder will have free rein, no matter how good the dial-in security is.

This is the prime reason why we say that security must be *global*. It is not sufficient to simply lock up part of the system; it has to be an all-or-nothing proposition, with cooperation, teamwork, and a consistent policy and commitment. Crackers are looking for a weak spot, a chink in the armor. A global security policy will deny them those weak spots.

Backups

Backups should be encrypted or at least protected by password. They should be kept locked, but accessible. Archive server software can be configured so that only an automated system can access the backup. This adds another layer of security. Remember the Maginot Line: Someone who reads your backup tapes has bypassed all your carefully designed LAN security.

Keep one set of backup tapes on-site and an earlier set off-site. There are published rotation schedules you can follow to assure that backups are done but do not bury you in tapes. Backups edge into the category of disaster recovery, and we have noted that security and disaster recovery are intertwined.[1]

Removable media

Many times access controls are placed on the hard disk. In DOS, these can be bypassed by placing a bootable disk in drive A: and booting from there. Then programs like Norton Disk Doctor can capture, read, and modify disk data, amend access privileges, or change security login routines.

Diskettes that are cleared of files still contain the old data unless magnetically erased, reformatted, or randomized. Disks containing sensitive information should be encrypted using DES. Then backups of these disks will automatically be secure.

Deliberately partitioned networks can be breached by a simple diskette carried from one machine to another. An insecure network can thus contaminate one with substantial security protection.

PCs, Macintoshes, and Notebooks

Machines should be physically secured if they are important. Various vendors make straps and wire cables that secure a PC to a desktop. Notebook machines can be secured with locks when in docking adapters. Do not embed password scripts in portable machines that can be stolen.

Transborder Data Flow

As an example of the extreme, international networks use many of the above facilities. General Electric Information Services' network spans 75 cities in 30 countries. They use both DES and ANSI encryption *and* proprietary message and user authentication tools. Such a system requires sophisticated key management, advanced cryptosystems, and constant vigilance by everyone involved. It also costs lots of money.

Keeping Perspective

Once again, we do not wish to fall victim to "Maginot Line thinking." The majority of risks come from *inside* the organization. Eyeball leaks are the best example.

As a people issue, *employee education* is the best weapon. Most people want to protect company data and they are the losers if it is compromised. Help them by keeping them aware of the need for security, so that it runs as a background program in their minds, all the time.

As one expert put it, "Most data thieves don't use super-advanced technologies. Instead, they sneak in through the back door." Help your people clamp those doors shut.

Review

- Network threats are basically interception, alteration, jamming, denial of service, or loss of integrity.
- LANs may be intercepted through network monitors or analyzers.
- Radio-frequency systems broadcast RF waves all over, making them easy to tap.
- Dial-in systems should be as low key as possible. Delay carrier tones. Give no information in the banner except a warning.
- Packet-switching network ubiquity has given hackers a rich hacking medium.
- There is some security in packet networks since individual packets can take different paths. This makes it hard for a hacker to copy an entire message.
- Leased lines, being "nailed-up," are fixed targets for tappers.
- Bridges are relatively insecure network *firewalls*.
- Routers are better *firewalls* to keep intruders out.
- Remove unnecessary protocols from routers.
- Gateways can be hacker targets if the hacker is knowledgeable.
- Removable media can contaminate even a secure network.
- Media should be erased to keep others from recovering files.
- In the end, the majority of leaks are internal. Usually they are unintentional, but there are those that are malicious.
- Employee education remains the best security tool you have.

Note

1 Regis J. (Bud) Bates, *Disaster Recovery Planning: Networks, Telecommunications, and Data Communications,* McGraw-Hill, New York, 1991; *Disaster Recovery for LANs: A Planning and Action Guide,* McGraw-Hill, New York, 1993.

Chapter

10

Firewalls: TCP/IP and Internet Security

"...it's safe to assume that if you typed your password on the Internet, someone else has it."

WILLIAM CHESWICK

Member of the Technical Staff
AT&T Bell Laboratories

Objectives

At the end of this chapter you will be able to:

- Present a cohesive case to management as to how and why or even if it is wise to provide Internet access
- Discuss the fields within TCP/IP headers that can be used as security "hooks"
- Understand why these fields are not foolproof and how they can be exploited
- Explain the principles, benefits, and drawbacks behind filtering
- Comprehend the problem filtering has with context
- Use firewall terms and synonyms accurately and properly
- Categorize the various types of firewalls
- Itemize risky protocols so they can be omitted from your firewall
- Appreciate and plan around router vulnerabilities

- Deploy packet filtering routers properly
- Deploy application gateways when their use is mandated
- Understand where circuit-level gateways are the best choice
- Evaluate firewall offerings completely and objectively
- Classify the basic firewall configurations and adapt them if necessary to meet your specific needs
- Analyze the vices and virtues of building your own firewall
- Comprehend the flaws inherent in TCP/IP
- Explain how IP "spoofing" works
- Specify OS weaknesses when using TCP/IP and how to eliminate them
- See how tunneling can permit a Trojan Horse attack
- Apply the SATAN (Security Administrator Tool for Analyzing Networks) and COURTNEY software tools to evaluate your own security's effectiveness
- Handle an incursion properly if the worst happens

TCP/IP and Internet Security

The Internet is the world's most fertile medium for spawning attacks against the networks connected to it. In security parlance the Internet is a *threat multiplier*, a system that multiplies the security risks already associated with networking. At the same time the lure of commercial opportunity and the temptation to mine its vast lode of information are enormous. From a security standpoint we are asked once again to find a balance of risk and reward. Helping you find the right balance of Internet connectivity and expressing that balance in implementable terms are the purposes of this chapter.

At one extreme, some network owners have imposed an absolute proscription on connecting to the Internet. This is unfortunate and largely unnecessary, but reflects the conservative attitude that any reasonable person would take without fully understanding the risks, rewards, and range of Internet security options that exist.

The other extreme is represented by groups that have full Internet connections yet no security protection at all. The National Computer Security Association estimates that less than 1 in 100 networks is protected.

While connecting to the Internet specifically poses some risks, it is not the only context into which a discussion of firewalls fits. If your organization has an internal network to which you wish to let others gain access, then you have an "inside–out" internet (with a lowercase "i") and the same problem as connecting to the Internet but in reverse. This chapter applies equally to you.

The Penalties of Excess Conservatism

Failure to utilize the Internet as a business medium places an organization at a competitive disadvantage in these ways:

- It precludes access to the Internet's tremendous information resources.
- It represents a lost opportunity to publicize products and services through an inexpensive sales medium.
- A firm loses the opportunity to shorten the sales cycle. Prospects can initially qualify themselves by reading and seeing a product on line. A salesperson can then call or see (since queries can be logged) this better-informed prospect and so reduce time wasted chasing dead ends. A World Wide Web (WWW) site represents proactive selling: Let your prospects come to you, rather than chase around looking for them.
- Loss of business may occur to competitors who describe their services and products on line, stimulating interest and predisposing prospective customers toward their solution rather than yours.
- The perception may arise that you are a hyperconservative, even regressive company that fails to keep up with the times. How do you answer when someone asks, "What's your e-mail address?"? A simple, direct answer is far preferable to anything else.
- A company fails to entertain the new ideas that discourse on the Internet inevitably stimulates.
- There is no opportunity to use the conveniences afforded Internet users, e-mail and file transfer to name just two. An Internet e-mail message costs about 6 cents. There is no cheaper form of electronic communication.

A matter of balance

For many organizations that would like to connect to the Internet, finding the right comfort level of connectivity is their core issue. That level of connectivity ranges from total openness to a totally closed system. Incredibly and as noted above, the National Computer Security Association estimates that less than 1 percent of companies connected to the Internet have some sort of firewall protection device in place. Put another way, 99 companies in 100 are totally open. No wonder then that we read so many stories of Internet break-ins in the mass media.

Categorizing the risks

Security problems in this environment can be divided into two categories: The first consists of the complexities of building restrictive devices to regulate traffic flowing onto and off of the Internet. Such devices are called *firewalls* and they are discussed below.

Innate flaws in TCP/IP that make security difficult are the second category. Some of these weaknesses are well known, making attacks easier to anticipate. Some of the major types of attacks are discussed toward the end of the chapter. You will also find some specific recommendations and references for securing TCP/IP networks insofar as it is possible. Many of the observations of TCP/IP's inherent security apply equally to other protocols such as Novell IPX and DECnet.

A Little Background on TCP/IP

One method of restricting the flow of information from one network to another is a technique called *filtering*. In filtering, a firewall device examines the content of a data packet or upper-layer control information, and sometimes even the content of the message itself. Knowing what is in this control information allows the device to make a decision as to whether to pass that unit of information on or not.

The control data that precede a sender's information are called the *header*. Each of these headers tells us something about the information that follows. When the message gets to the Internet Protocol or IP, the network layer IP header will indicate whether the message should be sent to the Transmission Control Protocol (TCP) or the User Datagram Protocol (UDP). Upon delivery to TCP or UDP, *port numbers* indicate the source of the application and its destination. Viewed in this way, there is a lot of information available for the looking.

Figure 10.1 illustrates a simple Ethernet frame containing an IP packet. The figure shows just what information is available in an

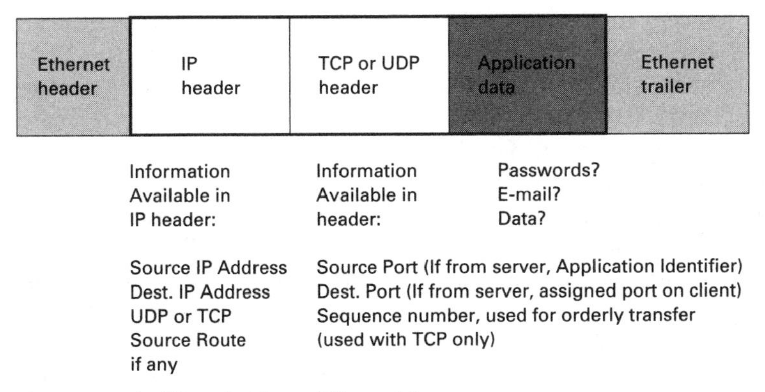

Figure 10.1 An IP packet in an Ethernet frame.

IP and TCP or UDP header that might aid an attacker or might aid in filtering out interlopers.

IP headers contain a source and destination IP address. These are the host numbers of the sender and receiver. Since Internet IP addresses are assigned by the Internet Network Information Center (InterNIC) and are public, it is easy to find out whose messages are being sent. Consistency in the higher-weight bits of the IP address can imply the use of subnet masking and allow an attacker to build a sketch of the target network's topology.

The Internet protocol allows *source routing*. Source routing specifies the path a packet must take to its destination. Accordingly, an attacker can use source routing to force a device to pass a packet to an intended target. For this reason, source routing must be turned off in firewall routers attached to an internet.

The Internet Control Message Protocol is installed with every IP implementation. By its nature, ICMP is inherently insecure. For instance, ICMP redirect messages can tell a host to send its packet to a different router. Falsifying such messages can cause packets to take a path to the attacker's machine.

Each IP packet must specify the kind of header that follows, either a TCP or UDP header. Since some applications use TCP (such as File Transfer Protocol) and others employ UDP (such as the Simple Network Management Protocol), an attacker can instantly learn the type of message in the packet.

Once the type of packet is known, the attacker can look in the TCP or UDP header and discover the exact application to which this packet pertains. This is possible because many applications in the TCP/IP suite are assigned *port numbers*. The first 1,023 port numbers are assigned by the Internet Assigned Numbers Authority and are available to anyone who cares to look. For instance, TELNET requests enter a TELNET server on port 23, Simple Mail Transfer Protocol on port 25, and Post Office Protocol (POP3) PC mail service is assigned port 110. All told, figuring out the source, destination, and content of a packet is easy.

TCP segments that follow the IP header also contain *sequence numbers*. Sequence numbers allow the receiving TCP software to detect missing, duplicated, or out-of-order segments. It is possible for a "spoofer" to guess some of these sequence numbers pretty easily as they often follow a predictable sequence in some Unix implementations. Using a combination of predictable sequences and a knowledge of the target's IP address, it is possible to prosecute an IP spoofing attack against a target. Spoofing software is freely available on the Internet. Indeed, this was the method allegedly used by Kevin Mitnick in an attack on Tsutomu Shimomura's computers on Christmas Day, 1994. A detailed discussion of IP spoofing appears toward the end of the chapter.

Filtering

These same fields can be made to work for us if we understand their limitations. IP addresses nominally identify senders and receivers, although these can be faked easily. On servers, port numbers equate to applications. Thus we can turn off requests from disallowed IP addresses and prohibit access to applications. We can even combine the two and permit or deny certain applications from certain IP addresses. One would think that we have a great deal of control.

This control is largely illusory because it is all too easy to falsify IP addresses and port numbers. Further, port numbers outside the range of 1 to 1,023 can vary widely when applications such as NFS, X-Windows, and NIS are used, making filtering rules more complex and thus more difficult to implement, verify, and maintain.

The Problem of Context

One of the difficulties with filtering is the matter of *context*. If I send you a letter as a complete entity, then you can read it and reply accordingly. However, if I shred it first and then mail it, you will have to painfully reconstruct the letter with much glue and patience. Trying to gain intelligence from a single shred is impossible, for in shredding the letter, I have made it nearly impossible for you to put that single shred back into its proper place in relation to other small pieces of information in the letter. I have removed its context, and probably lost a friend in the process.

In networking, we shred, or more properly, segment, information all the time. We take e-mail messages and complete files and break them up into small pieces called TCP segments, typically 536 octets long or less. Each TCP segment becomes a single IP packet. It may take dozens, hundreds, or thousands of packets to complete an entire message or a file being transferred. When we do this segmentation, we lose context.

Trying to make allow–disallow decisions based on small pieces of information is difficult enough in TCP sessions, but the use of TCP port numbers (which equate to TCP-based applications) gives us some sense of continuity. Sequence numbers used by TCP to maintain an orderly flow of traffic give us additional coherence. However, the enormous variety of UDP port numbers used and their arbitrary selection when used with X-Windows systems and remote procedure call (RPC) portmappers make the recovery of context in UDP sessions essentially impossible. Thus the conventional wisdom says that X-Windows systems and systems using RPCs, NFS, and NIS cannot be effectively regulated by use of packet filtering.

Some packet-filtering vendors restore context by rebuilding applications using the same port numbers. This is because although arbitrarily chosen, once a port number is selected, it stays the same for a session. With the appropriate intelligence, a firewall can, in effect, unshred the document provided that it can reassemble context using port numbers, application information, and the limited information in each unit of UDP control data.

Firewalls

Many people have heard or believe that a firewall prevents unauthorized and improper transit of access and information from one network to another. The choice of the term *firewall* is unfortunate, because it leads one to think of a simple, unbreachable barrier. This notion goes beyond simplistic—it is genuinely naive and even dangerous to think of any firewall in this way.

Firewalls do indeed provide a degree of isolation between two networks. Properly chosen, configured, and maintained, they will provide a specified level of security. In this chapter we offer several firewalling techniques, itemize their pros and cons, and help you choose the technique best suited for your need.

Firewall terminology

A discussion of firewalls can be confusing due to the use of loose and similar-sounding terminology. Likewise, some of the physical configurations can be confusing because although they may *look* much the same, they *work* differently. The major terms used in this chapter are defined below.

A *host-based firewall* refers to a general-purpose computer in which security software is running. Such machines examine activity at the application level rather than at the network level. The software is referred to as an *application proxy* and usually consists of a client and server to service both sides of the gateway.

Because they are configured to resist attack, host-based firewalls are also called *bastion* or *gateway hosts*. More specifically, *proxy servers,* also called *application gateways,* are examples of host-based firewalls. We use the term application gateway since the term is more descriptive of what it does.

A *router-based firewall* uses routers to screen TCP/IP headers as a basis for blocking or permitting traffic at the network level. Such routers are called *screening routers* or *packet-filtering routers*. Here we use the latter term.

Dual-homing is a TCP/IP term that refers to a device with more than one IP address. Therefore a dual-homed gateway consists of an application gateway with two network interfaces, one on the protected network and the other on the Internet or nonsecure network. IP forwarding is disabled and the machine runs proxy software only for specifically allowed applications. A dual-homed gateway is the most secure form of firewall.

For this discussion, a *nonsecure* network is a broader term than the Internet. While most firewalls are barriers to the Internet, they can also be used as barriers to other networks as well. To make that point, we use the terms interchangeably.

All or nothing

Some firewall configurations have an inherent weakness. If an attacker succeeds in piercing a single firewall, notably a packet-filtering firewall, they then have total access to your system. This violates our perimeter rule in which we use increasingly stronger defenses as we approach the core information or resource sought. We use attrition to filter out, in stages, those who do not have permission to go any higher. Indirectly we also establish security levels appropriate to our user community.

The all-or-nothing nature of a single firewall is solved by the use of several firewall devices in tandem. Dual-homed gateway firewalls, screened host firewalls, and screened subnet firewalls all have multiple barriers to eliminate the all-or-nothing aspect provided by a single firewall.

Firewalls are classified into the following subcategories:

- Passive firewalls using intermediate networks
- Packet filters
- Circuit-level gateways
- Application gateways

We will discuss and compare all four techniques. Often, more than one type is used at a time to multiply the security's effectiveness.

Intermediate Networks as Passive Firewalls

Sometimes using an intermediate network is a sufficient insulator from the Internet itself. For instance, using Internet services through Prodigy Services, America OnLine, MCI Mail, SprintMail, or AT&TMail is fairly safe. While it is certainly possible to download a file containing a virus, the ability of an attacker on the Internet to gain access to your machine that is using proprietary non-TCP/IP software is very limited.

Much of the insecurity on the Internet stems from the ease of manipulating TCP/IP, the flexibility of Unix hosts, and deliberate social engineering attacks. Proprietary network users are immune from the first two but not from the third. Give out no personal information and scan every file that is downloaded for viruses, whether the file is executable or not.

Packet Filters

As we have discussed, the concept behind packet filtering is simple. Just have a router look at the TCP/IP headers. If you see something you do not choose to allow, be it an application or an unknown or undesired IP address, throw it away. Place the router between, say, the Internet and your internal network, and you now have a *firewall* between them. You think you are secure; *but you are not.*

By now you know that the widespread belief that firewalls are *the* solution to TCP/IP network security is naive. Attackers will exploit subtleties that you may not realize exist in an effort to gain illegitimate entry. We cannot afford to look at firewalling at only the gross level and remain blind to the tiny details that attackers will take advantage of in order to gain the keys to your kingdom.

It is important to understand several things:

- Routers were never designed to provide security. Their purpose is just the opposite: to facilitate communication, not block it.
- Security is a bolt-on addition to TCP/IP. It can never be as good as a security mechanism woven into the fabric of a protocol suite.
- Proper packet filtering is difficult, detail-oriented, and fragile. It is contrary to the nature of corporate life in which everything is rushed; minimal time is taken to think things through and no one wants to be mired in detail.
- The fragility of firewalling means that every system change, even an update to the router software, calls for a reassessment of the firewall programming in place. Automated tools for this purpose are almost nonexistent, forcing time-consuming and incomplete testing by hand.
- Once in place, even the best firewall can never, ever be taken for granted.
- Even if your application environment is frozen in place, it does not change the fact that the threat is constantly evolving. What is adequate security today is marginal tomorrow and inadequate the day after. Complacency can be our undoing; unending vigilance is therefore the price of security.

This is not to say that firewalls are ineffective and do not provide a reasonable or even high degree of security. However, some sophisticated users such as General Electric have been successfully attacked (some of GE's passwords were stolen). A client of one of the author's friends insisted that he did not need a firewall before connecting to the Internet. He did so without installing a firewall. A week later his firm was infected with a virus. (Hopefully you will object to this last statement. If you have read Chapter 2, you know that most viruses travel by diskette. However, the client did open his network to another source of infection.)

Even given the existence of Internet firewalls, a poll of managers at one security conference revealed that a little over half of them remained restive about their Internet connection.

Packet filters, when installed on routers, turn the routers into firewall routers. Firewall routers should do no routing, but sometimes do anyway. As discussed above, packets not explicitly allowed are discarded. Filtering can take place in your own firewall router, your Internet network service provider's router connection to you, or both.

Fields alone in the TCP/IP protocol are inadequate to provide the level of resolution often needed, e.g., "allow *this* but not *that*" within an application. For instance, while it is possible to prohibit or allow File Transfer Protocol, it is not possible to allow the GET function of FTP while disallowing the PUT function when filtering is used. The entire protocol must be allowed or not. Filtering TCP segments is difficult, but doing the same for User Datagram Protocol messages is nearly impossible because UDP offers essentially no context.

Packet filters do not cope well with FTP or Domain Name Service. For example, it is easy to falsify an outgoing message into an allowed form to get through the filter. In sum, packet-filtering firewalls are good at looking at routing information, but they cannot see into the application where the need and control possibilities are greatest. Being near-sighted at higher layers, packet-filter firewalls depend on secure implementations of those layers. Insecure implementations compound the security problem.

Filter tables

The work involved in setting up filter tables is complicated, and each router vendor's syntax and format are different. If the tables do not cover every possible condition, then the firewall has a security hole in it. There is also a small negative effect on performance as the router does more processing of the TCP/IP headers than it would normally do.

The filter setup process has three steps:

- Decide what you will and will not allow. This flows directly from your security policy.
- Specify the permitted packets as logical expressions.
- Rewrite those expressions in whatever syntax your router vendor uses.

As Fig. 10.2 shows, it may not be necessary to isolate your network totally from the outside. It may be wiser to segregate your network into that part which can communicate freely and that part

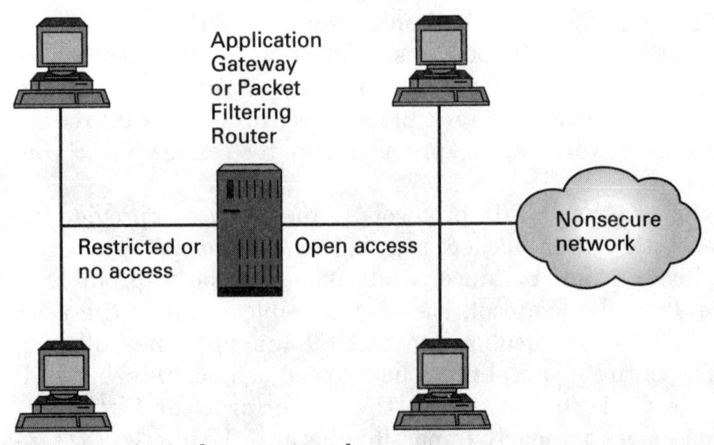

Figure 10.2 A mixed-access network.

which requires isolation. Because the latter may be hard to specify some people simply choose to isolate their network altogether. However, the benefit of having such a mixed-access network is that the unsecure part is easier to use. This is consistent with the security balance we seek.

Firewalls should disallow services that are not needed or those that could cause compromise. They include Network Information and File Service and the unauthenticated remote "r commands" such as rsh, rcp, and finger. Remove, block, or restrict protocols from the router that might help a attacker.

Remove or block these protocols:

- *X-Windows* and *OpenWindows* (ports 6000 and up and port 2000, respectively) place display data and keystrokes on the network.
- *TFTP* (port 69), the Trivial File Transfer Protocol, moves files. TFTP can be used to update router software. Do not let an attacker update yours.
- *RLOGIN* is remote login (port 513); *rsh* (port 514) is for remote shell; *rcp* for remote copy, *rxe* (port 512) for remote execution. R commands have no passwords, and if improperly configured allow access to unauthorized accounts and commands.
- *RPC,* remote procedure calls (port 111) include NIS and NFS. These protocols can be used to steal passwords and files.

These protocols are usually allowed but restricted:

- *Ping* (port 7) is a little echo program to test continuity and path delay. It allows an attacker to see if a host is reachable. Do not give an attacker the use of your tool.
- *FINGER* (port 79) shows exactly who is logged into a system on a LAN and from where the login originated.
- *TELNET* (port 23) allows clients and servers of many types to speak, including attackers to your router.
- *FTP* (port 21), the File Transfer Protocol, copies files in ASCII or binary and allows housekeeping such as renaming or deleting files.
- *SMTP* (port 25), the Simple Mail Transfer Protocol, is often restricted to a single e-mail server.
- *RIP* (port 520), the Routing Information Protocol, can be spoofed.
- *DNS* (port 53), the Domain Name Service, contains host names and host information; can be spoofed.
- *UUCP* (port 540), Unix-to-Unix CoPy, can give unauthorized access if improperly configured.
- *NNTP* (port 119), the Network News Transfer Protocol, assesses and reads news.
- *GOPHER* and *HTTP* (hypertext transport protocol; HTTP is on port 80) servers should appear on proxy servers only.

Router modifications

All communication in TCP/IP requires at least half duplex capability. You can modify routing tables to disallow a route back to the originator and so prevent a connection. While a rather brute-force method, it is simple and it works. Similarly, routing protocols should not be started at boot time, and no default routing protocols should be present. Host-specific routes are all right.

Check inbound remote users against an IP-address access control list. Disallow any activity until this process is completed satisfactorily. During off-hours, the IP routing protocols should be turned off. During regular business hours, permit only known source and destination addresses to pass.

Ironically, the presence of routers gives rise to a new kind of attacker—*router hackers*. They watch packet addresses and data.

They can corrupt routing tables and metric values and reroute packets anywhere they like. They can add or delete packets. The venerable TCP/IP Routing Information Protocol contains *no mechanism to authenticate a routing table update*. However, some proprietary protocols do, as does the Open Shortest Path First routing protocol. Routers can be accessed via SNMP and can sometimes be accessed via TELNET. Router updates can even be sent using the Trivial File Transfer Protocol. Any of these protocols that are used should be fully authenticated and protected by encrypted password.

Router address tables contain network numbers and sometimes device addresses. The tables could be used to jump from one network to another or to make a router think an intruder is on an approved network segment. To prevent this kind of problem, encrypt the routing tables and authenticate them so the router cannot be fed false information.

All of these represent real vulnerabilities. Ask your router vendor(s) for recommendations on how to make these machines more secure. Some routers offer encryption, either of an entire packet or leaving the headers in plain text. Two of these techniques will be discussed later. Routers connected to the Internet need such special attention, especially in view of the statistics quoted at the beginning of this chapter.

Packet-Filter Firewall Advantages and Disadvantages

Packet-filter firewalls are erected to keep the Internet vandals out of our corporate networks. They are not all they are cracked up to be (the firewalls, not the vandals).

A summary of the advantages and disadvantages of packet-filter firewalling appears in Table 10.1. As we have seen, packet-filter firewalls require a great deal of careful configuration. Incorrect programming can create unintended security holes. Router software updates, which try to preserve such programming, may not always be able to do so.

When is packet filtering alone enough?

Packet filtering alone may be adequate if:

- Your organization is very decentralized.
- There are many ways around your firewall.

TABLE 10.1 Advantages and Disadvantages of Packet-Filter Firewalls

Advantages	Disadvantages
Reduces the risk of security breaches	Vendor software upgrades can degrade the effectiveness of filtering
Reduces time spent coping with the result of intrusions: viruses, corrupted data, stolen files	If short on Internet savvy, a conservative approach is necessary
Reduces cost to and potential disruption of the business	Desirable or convenient services can become unavailable due to the context problem in filtering
Minimizes legal and financial exposure	Complexity reduces security
Avoids the irony of implicitly sponsoring attackers by giving them access to your computing resources	Hardware, software, and maintenance costs
Allows access to Internet resources and business opportunities	Software maintenance and upgrade costs
Is one answer to Internet access versus the need for privacy	Initial setup and ongoing administrative costs
If well executed, gives attackers nervous breakdowns and sends them somewhere else	Lost business due to a broken or malfunctioning firewall

- Your organization has no strong central security policy.
- The number of hosts in your network is very small.
- You rely heavily on host security to keep attackers out. Remember, however, that host security does not scale well. As the number of machines increases, the odds favor a hole somewhere.
- You use non-Unix servers such as Novell NetWare and workstations based on DOS/Windows or other non-Unix operating systems (a few Unix machines are all right provided all the usual security holes have been plugged).
- You do not use dynamic IP address assignments such as Dynamic Host Configuration Protocol.

When is packet filtering alone inadequate?

Packet filtering alone is inadequate if you have a range of Unix, MVS, or VMS machines over which you have no security control.

To regain control, a stronger firewall is needed, often a dual-homed machine as described below.

A firewall or packet-filtering router is best used only as a firewall. Today, an assortment of vendors make firewall routers that are specifically designed to keep the attackers at bay. See the end of the chapter for a list of vendors in this market.

Always keep in mind that *your security is only as good as the eyes on your network.* Those eyes consist of packet-filter firewalls and gateways designed to inhibit improper traffic, plus a real-time network monitor with appropriate logic traps.

Application and Circuit-Level Gateway Firewalls

Gateway firewalls fall into two categories: circuit-level gateways and application gateways. We discuss application gateways first.

Application gateway firewalls

Application gateway firewalls are just the reverse of a router firewall. Whereas router firewalls are general purpose and regulate all the different kinds of traffic we have discussed, application gateways use specially written code for specific applications. These sets of code are called *proxy services* and exist both as clients and servers within the physical application gateway. All services are passed between clients and servers in the gateway. If you look at Fig. 10.3, which illustrates a dual-homed gateway, you will see the logical construction of an application gateway.

If based on Unix, the application gateway will have several host daemons (processes), one relay daemon for TELNET, another for FTP, perhaps. Client software connects to it. The relay daemon sets up a connection to the Internet and to a server site.

In order to make this approach feasible, it is necessary to have a situation in which only a few programs need to be ported from one side to another because the complexity builds quickly as applications are added. From a practical standpoint, there are not that many client and server software packages around. Unless you choose to write your own, your choices are limited to about ten more-or-less standard applications such as TELNET, FTP, and SMTP. Any applications unique to your site require custom-written proxy services specific to the applications involved. This limits flexibility as new services are added.

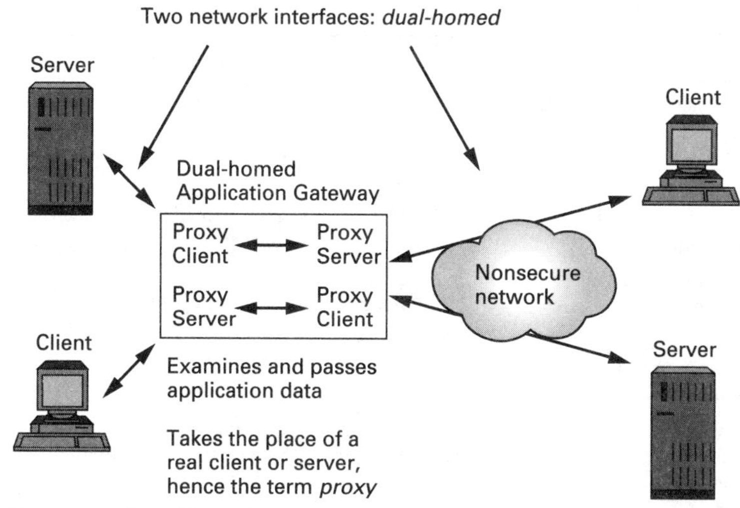

Figure 10.3 A dual-homed gateway.

Having better and specifically designed hooks, application gateways are simpler and hence more secure. The software hooks built into the application make it easy to control all incoming and outgoing traffic. Application gateways are often used with electronic mail, for instance.

The negative side of the application gateway approach is almost self-evident from the description above. An application gateway requires a specialized user program or a variant of the user's client software as a hook on which security decisions can be made. The word "special" implies extra work, a certain degree of added complexity, and a slight sense of exploring the unknown. For application gateways, all of these implications hold true. (See Table 10.2 for a list of advantages and disadvantages.)

Such special effort cannot be justified for dozens of applications: Only a few can receive such attention. In a very heterogeneous shop, the application gateway approach becomes unwieldy and too laden with overhead (in every sense) to be workable. The choice then devolves down to whether to use the application at all; take the risk and use it anyway or spring for the necessary special support. When a new application arrives on the scene, the choices are much the same, making it more difficult to integrate a new technology.

Since the investment value disappears when an old application goes away, part of the support decision is the expected lifetime of the application. This can be singularly hard to judge and IS man-

TABLE 10.2 Advantages and Disadvantages of Application Gateways

Advantages	Disadvantages
Allows only services for which a proxy exists	A two-step process for client–server applications: Connect to server gateway; gateway connects as client to external server (and vice versa)
The protocol itself can be filtered (e.g., PUT command in FTP)	May require modified clients (oddly, this can add to security)
Hides host names and IP addresses; outsiders see only the gateway	May force users to change their normal work pattern by adding steps
Excellent authentication: advanced authentication is possible from outside hosts using one-time passwords	More expensive than a packet-filtering router
Allows authentication to reside only on the gateway; internal host security becomes less important	Changes to applications require changes to the gateway; less flexible than a packet-filtering router
Superior logging capability	Requires all supported services, called *proxy services*
Simple filtering rules: Send all traffic to the gateway and reject the rest	Difficult to adapt to new services such as World Wide Web and Gopher
Often used for TELNET, FTP, e-mail, X-Windows, plus other services	Investment value in the gateway is limited by the lifetime of the applications it runs
Gateway can check source IP addresses	
Centralizes mail collection and distribution	

agers have been known to consult several crystal balls of varying reliability before making such decisions.

Gateways allow communication with the gateway and no farther. They are generally preferred from a security standpoint to firewall routers, being easier to configure and thus more secure. However, a gateway must often be capable of supporting or at least mimicking an application. For instance, an application gateway firewall may also be asked to function as a domain name or mail server.

A gateway firewall provides no information about internal host IP addresses or names to the external network. Aliases to the outer world are used and the firewall passes these to an internal name server for translation. Thus the firewall appears as the originator and recipient to the outside world. True aliasing requires that message headers be stripped of IP addresses and host names when going outbound.

Application gateways resolve the problem of the network manager who has used a block of IP addresses internally but did not register them with the InterNIC. Since application gateways hide internal IP addresses from the Internet, only a few registered addresses are required. Internal renumbering becomes unnecessary.

Two domain servers are required for use with an application gateway: One outside the secured network that claims to be authoritative for the network to the Internet and another server on the secured network that is truly authoritative. There is some disagreement among experts as to whether having two servers is worth the trouble. In the author's view, anything that denies an intruder information is worth the effort.

Application service often includes file transfer protocol service as well. Internal users log into the firewall to transfer files in and out, as do external users. If anonymous file transfer is allowed, it is permitted only to and from the firewall. The same applies to TELNET or rlogin sessions: The user TELNETs to the firewall; the firewall then does so to the outside. Electronic mail, FTP, outside logins, and services offered uniquely by the site are most safely supplied this way. The gateway itself can then connect approved outside users to internal services, perhaps using Kerberos client–server security to manage the session.

It is common to use application gateways in conjunction with the packet-filtering firewall routers discussed earlier, as well as the circuit-level gateway discussed next.

SEAL (Screening External Access Link) DEC's SEAL is an example of a high-level application gateway firewall. SEAL is software that resides on a dedicated network server that is placed between an internal network and the Internet. Internal users can gain access to the Internet, but external attempts at access to the internal network are limited by what the internal network owner wishes to allow.

SEAL logs all access, both incoming and outgoing. Access is allowed based on authentication of user identification. Outbound, SEAL acts as a transparent relay that intercepts and processes commands from the internal user. Accordingly, all IP packets stop at the gateway. SEAL then regenerates that information in the form of a service such as e-mail or FTP onto the Internet.

While only one server can be equipped with SEAL, the server that scans packets and the one that relays the application commands can be two separate devices. This is the way DEC has their own SEAL gateways configured.

For example, when a DEC employee wishes to copy a file from the Internet via FTP, the employee's request is forwarded to SEAL. Through the gateway, SEAL obtains the file, scans it, and regenerates it into FTP as a server to the user's FTP client software. SEAL can also work with other services such as World Wide Web.

Access to the Internet is not limited to the workplace. Working from home, users can authenticate themselves to SEAL through the company's internal network (via dial-up, for instance) using a cryptographic key in the form of a "smart card." The cards are manufactured by Digital Pathways. Security Dynamics provides the authentication technology. Using the latter, random passwords are used to validate network users from outside the office. Once validated, dial-in users can perform third-party transfers of files from inside the organization to an outside host in safety.

To properly implement SEAL, DEC offers security consulting to ensure the product will work as desired. The consulting phase consists of developing an Internet security policy, definition of access rules, installation and configuration of any customized software, and training in the operation of the gateway.

SEAL's hardware platform is a DEC computer running Ultrix (DEC's version of Unix) or DEC OSF/1 (the Open Software Foundation's version of Unix). Software is available that supports e-mail, FTP, TELNET, information services, and conferencing. SEAL detects probing attempts and allows auditing of access attempts. As a high-level product, SEAL is expensive, with prices starting at $25,000 and up. It is presently available in the United States, Canada, Latin America, Europe, and Asia.

Circuit-level gateway firewalls

A distant cousin to the application gateway, the circuit-level gateway firewall simply relays TCP connections as shown in Fig. 10.4 such as an outside host connecting to an internal host. If used often, the process can be made automatic. Otherwise the caller must specify the desired internal destination and service. The gateway either makes the connection, or if disallowed, responds with an error code.

Circuit-level gateways are sometimes preferred if most of the traffic is outbound. In this mode, the caller connects to a TCP port on the gateway. The gateway then connects to the destination. Once the connection is established, a relay program copies data moving in either direction. SOCKS (for SOCK-et-S, a development name that stuck) is a generic circuit-level gateway that relays TCP segments and is discussed later in this chapter.

One problem with circuit-level gateway firewalls is that, like their application gateway cousins, a new internal program or resource will require code changes in many circuit-level gateways (SOCKS is an exception).

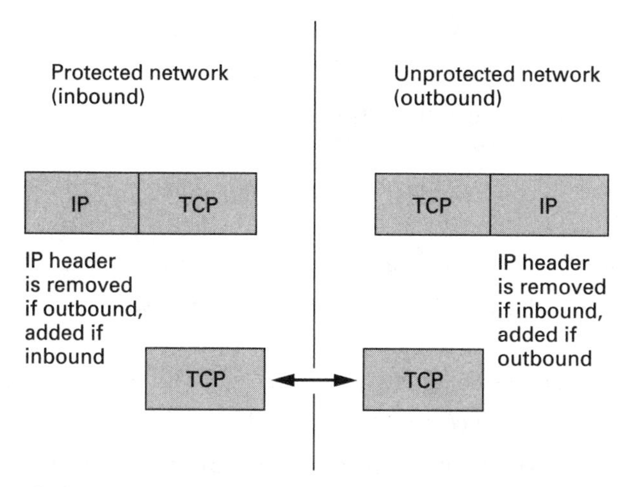

Figure 10.4 A circuit-level gateway.

General Gateway Notes

Gateways can also perform filtering. As full seven-layer devices, gateways do more than bridges or routers and are fertile ground for attackers. For instance, if an attacker can control the layer-5 session controls, they can let a user log into a resource and wait for the user to log out. The attacker can then kill that command and take over the session.

Networks should not leave sessions open. If a session has been inactive for a period of time, it should be closed out by the system. In some implementations, TCP sessions will be closed after two seconds of inactivity. If an external router contains a path to an internal location other than the gateway, then by definition, there is a leak.

Firewall Buyer's Checklist

General criteria

Here are some general criteria, adapted from NIST 800-10 (see the Appendix for where to obtain a copy of this valuable document), followed by some specifics:

- The firewall should implicitly deny services except those explicitly permitted.

- The firewall should support your policy, not impose one.
- Flexibility toward new services and policies is important.
- If an application gateway, it must support the proxy services you need.
- Simple Mail Transfer Protocol should be supported to centralize mail handling.
- Public site access should be supported to information servers.
- The firewall should be able to concentrate and filter dial-in access.
- If the firewall uses an operating system, it must be secure.
- The firewall's strength and correctness should be verifiable.
- Updates and patches should be offered periodically and installed promptly.
- Adaptability is important.

Specifics

When evaluating firewalls, look for the following:

- What *is* the device you are evaluating? Is it a packet filter, circuit, or application gateway?
- Protocol support (this includes data link, network, and application protocols).
- Ease of configuration.
- Ease of management.
- Logging and database manipulation/exportability for data reduction.
- Intrusiveness into user habits and operations.
- Advanced authentication capability such as use of security tokens.
- Simplicity of its management interface (does the interface make it easy to understand what you are doing?).
- SNMP interface to a network management system.
- Filtering flexibility including bidirectional filtering.
- Simplicity and power of the vendor's filtering syntax.
- Dual Ethernet connections to isolate an internal network from an externally accessible server (dual-homing).

- Availability of internal encryption and authentication of headers, contents, or both.
- Time of day restriction.
- User restriction.
- Availability of an evaluation unit (essential).
- Telephone, network, and/or on-site technical support: Quality, availability, responsiveness, and depth of knowledge.
- Warranty terms.
- Maintenance contract terms and price.
- Upgrade terms and price (e.g., software, firmware, hardware).
- Purchase, lease, or lease–purchase terms (the author strongly favors leasing these devices as they can become obsolete quickly).
- Statefulness of the device: Can it track individual connections and so gain a contextual sense of the information being sent in order to make allow–disallow decisions?
- Is the device a complete unit, or is it software running on a generic piece of hardware? Who will be responsible when the machine misbehaves?

Configurations

Having introduced the principles underlying the packet-filter and application gateways, it is time to examine some ways to use them, either alone or in combination, and explore their benefits and drawbacks. In increasing order of security, the scale is as follows:

1. Packet-filtering firewalls simply filter packets.
2. Screened-host firewalls allow trusted services to bypass the gateway.
3. Screened-subnet firewalls can be nearly as secure as dual-homed gateways.
4. Dual-homed gateways allow absolutely no access to internal systems.

The following descriptions are important because they express the logical relationships between machines, whereas figures tend to focus on physical connections. It is the logical relationships, the way the machines see each other, that is important to understand.

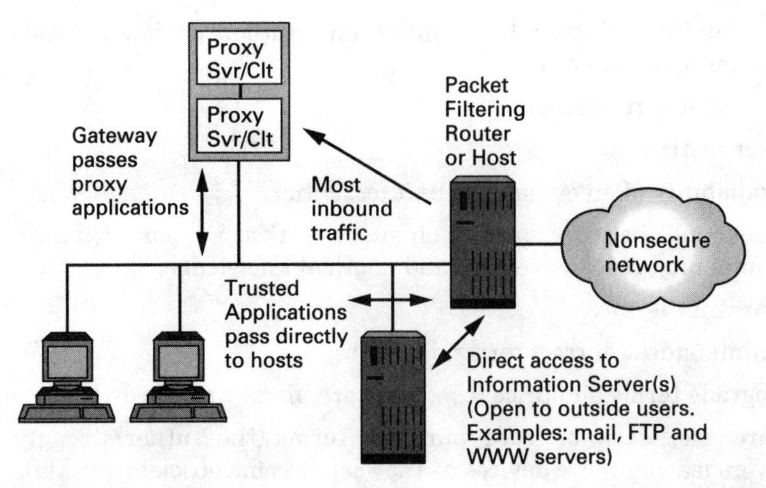

Figure 10.5 Logical routing.

There are many variations in these logical relationships. For instance (see Fig. 10.5) some sites pass all inbound traffic to an information server through an application gateway, while others pass the traffic directly to the information server. Such a decision depends on whether gateway software exists or must be written and the assessment of the risk involved in allowing direct access to the information server. If constructed as a bastion host, then the risk is low.

For all subnetworks used below, employ a class C address to limit the number of possible IP addresses that can be targeted.

The Packet-Filtering Firewall

As Fig. 10.6 shows, a packet-filtering firewall is a simple and common way of providing minimal Internet security. The filter can be installed at the Internet connection or on any subnetwork, leaving the rest of the network completely open. Systems on the "protected" side of the firewall have direct access to the Internet but inbound access from the Internet is mostly blocked, especially risky services such as NFS, X-Windows, and NIS. The advantages and disadvantages of such firewalls are tabulated in Table 10.3.

The Dual-Homed Gateway

A variant of the application gateway, the dual-homed gateway is much superior to the simple packet-filtering router. "Dual-homing"

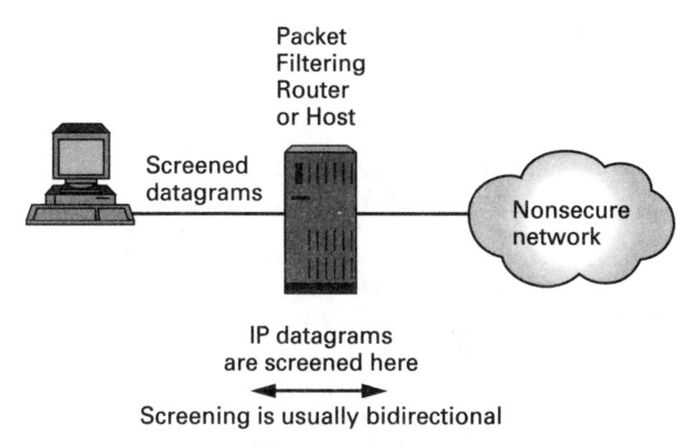

Figure 10.6 A packet-filtering firewall.

TABLE 10.3 Packet-Filtering Firewalls

Advantages	Disadvantages
Easy to employ; inexpensive	Not very scalable as security needs increase and become more complex
For small, uncomplicated sites	Little or no logging
May examine network port of origin	Hard to test rules implementations, manually at best
Specific connections may be allowed to specific hosts; connection types may be different by host	Rules become complex quickly, reducing overall security
Can block connections to or from specific hosts, networks, or ports such as untrusted source IP addresses (or even block all external addresses)	Hosts accessible from the Internet require their own copy of advanced authentication software, an expensive and complex proposition
Difficult to implement the usual rule: "Deny all except that which is explicitly permitted"	May not filter source UDP or TCP ports
If the router has two or more interfaces, testing the network interface on which a packet arrives and the one to which it is bound is a real advantage. This allows the router to filter both in and outbound, and permits the detection of spoofing attacks	Exceptions are hard to implement and reduce security
An all-or-nothing approach	Hard to filter RPC services due to dynamic portmapping in RPC protocol

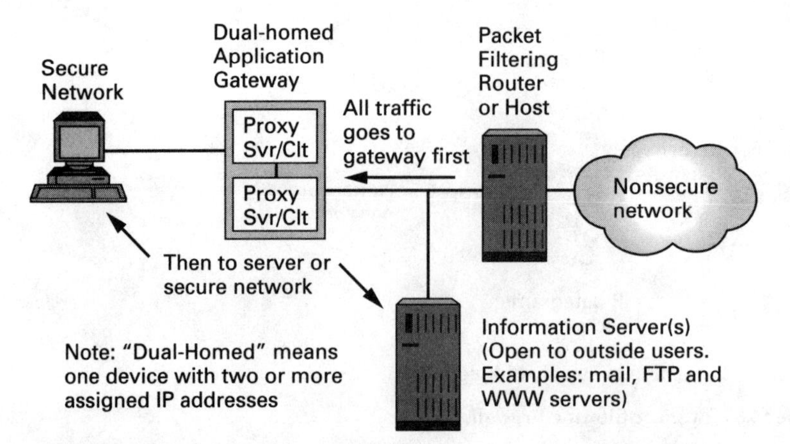

Figure 10.7 A dual-homed gateway firewall.

is a TCP/IP term that refers to a host with more than one IP address. In this case, each network interface has its own IP address as shown in Fig. 10.7.

The host cannot forward packets from one network interface to another, but can forward packets to open information servers such as e-mail, Gopher, or WWW machines. It cannot accept or deliver source-routed packets. The gateway host can authenticate inbound or outbound users and pass proxy requests and responses from secure internal machines. The network between the packet-filtering router and the gateway where open machines can reside is called a *screened subnet*. Because of its two-stage security, the dual-homed gateway is very secure and not very complicated to set up and manage.

As in the application gateway, internal network names and numbers are completely private and cannot pass through the gateway. Some gateways ensure that e-mail headers containing IP addresses and real host names, for instance, do not accidentally pass through the gateway outbound. Instead, aliases are used inbound. No DNS information is passed.

One benefit of this scheme is that a certain number of services can be public yet safe. The information server is an electronic market for external users who seek information on products, access to Web services, and so on. These information servers are also protected because the gateway acts as a proxy to the outside requester and acts in lieu of the requester to the information server. Even if the server is penetrated, the gateway prevents further access. On the other hand, some users elect to make the information server directly accessible by the outside user, thus enabling its IP address and name to be listed by public DNS servers.

The dual-homed gateway must be very secure, with all unnecessary and compromising services turned off, such as IP forwarding. Special TCP applications act as proxy agents. When a user TELNETs to the gateway, the proxy agent asks for the Internet address sought, not an input prompt as is usual. The TELNET proxy agent daemon then sets up the connection, which it can then monitor and control. In this way, individual commands within an application can be allowed or not, as desired. A packet filter either allows an entire application or not, and so is far less granular.

Dual-homed gateways are very secure but can be a bottleneck and a single point of failure. Upgrading them to faster machines all the time is not a good strategy. Either overbuild the machine or (less desirably) add gateways in parallel. Due to the special software required, adaptability to special needs is not the gateway's forte. The next two configurations add the flexibility that the dual-homed gateway lacks.

The Screened-Host Firewall

At some expense to security, flexibility is gained by using a *screened-host firewall,* shown in Fig. 10.8. Because "trusted" inbound applications can pass directly to the internal network after passing through the packet filter, the potential for a security breach is opened. If proxy services do not exist or if the risk of using such services has been evaluated and found acceptable, then these applications may be considered "trusted." Applications such as Network Time Protocol, which is considered low-risk, could qualify. It is also fairly common to allow Domain Name Service so that hosts on the inside of the packet filter can access Internet services.

While a packet-filtering router is used here, its rules definitions are not as complex as if a packet filter were used alone because all nontrusted application traffic is sent to the gateway. There must also be concern for "trusted" applications. Since trusted applications bypass the gateway, the inducement to compromise a trusted application exists, even if it is hard to subvert. Note that such applications have direct internal network access. An insider could attack a trusted application, and if they succeeded, you would have no way of knowing it.

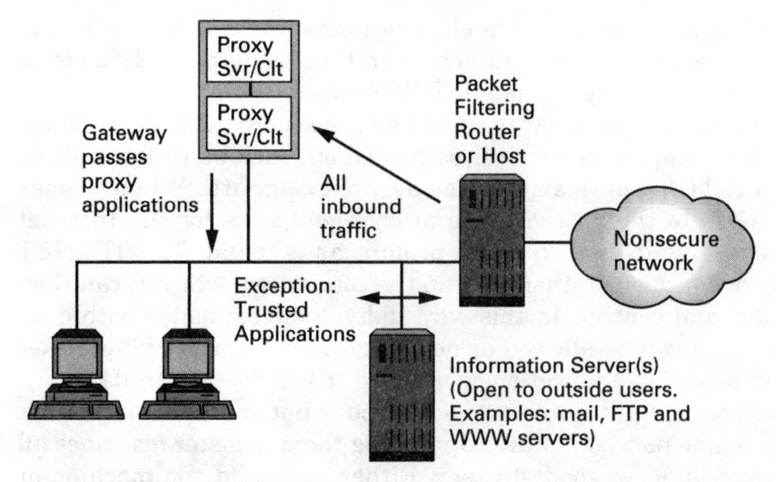

Figure 10.8 A screened-host firewall.

One benefit of the screened-host configuration is that only a single application gateway network interface is needed. Inbound traffic destined for the application gateway is passed on through the packet filter. Anything else is rejected. Outbound, the router rejects any traffic that did not come from the application gateway.

The above assumes that the gateway is always in the serial logical path, whether traffic is going to internal hosts or to the information server(s), if any. The screened-host configuration may be set up to route traffic directed at the information gateway directly to it, bypassing the application gateway. In this case the router would have three inbound targets: the gateway, the information server, or an inside host receiving a trusted application. All the devices could pass traffic outbound through the router.

The Screened-Subnet Firewall

A screened subnet is shown in Fig. 10.9. It adds another packet-filtering router to isolate the internal network further. The network connecting the two routers, the gateway, and any information servers is called a *screened subnet.* Modem pools (as either LAN-compatible network appliances, communications servers, or terminal servers) could be placed here, although it is generally preferred to consider them nonsecure and place them outside the packet-filtering router. This means that dial-in users are treated with the same suspicion and care as inquiries from the nonsecure network, e.g., the Internet. Modem pools are discussed further below.

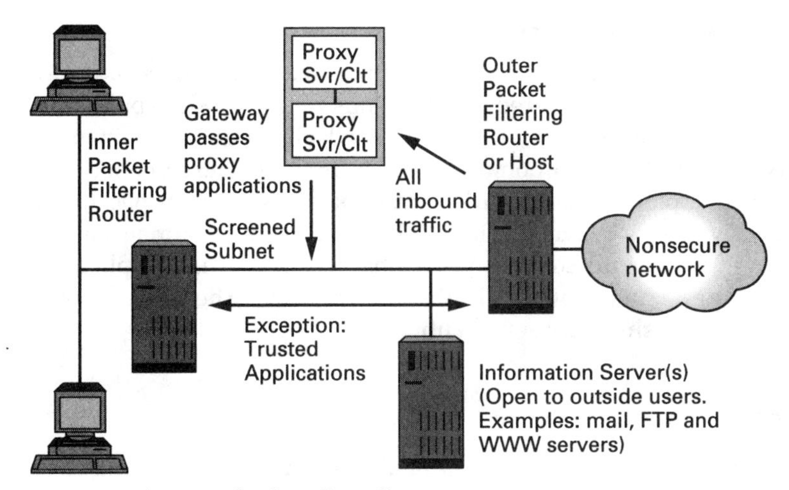

Figure 10.9 A screened-subnet firewall.

The outer router

Inbound, the packet-filtering router or host sends electronic mail and application traffic to the gateway alone (e-mail proxy service is often found on application gateways). FTP, WWW, and other such information inquiries may go to the application gateway first and then to the information server or directly as discussed earlier. Any other inbound traffic is rejected.

Outbound, the paths are just reversed. Outbound traffic to the Internet is routed, but any traffic targeted at an inside host is rejected. Further, the router rejects any traffic from devices that should not originate connections, including the information server, site hosts, or communications servers. Insecure protocols such as NFS and NIS are also blocked.

The inner router

Inbound traffic from the application gateway (including e-mail) passes through the inner router. All other traffic such as FTP and WWW service was targeted by the outer router at the application gateway, which then targeted the appropriate information server. Accordingly, no such traffic should be knocking on the inner router's door. Even if it does, perhaps due to misconfiguration, it will be rejected. Outbound traffic is treated the same way and is targeted solely at the application gateway. Note that outbound FTP, WWW, and similar requests will be honored whether to the Internet or the local information server. Dial-out modem pool calls could also be accepted if desired.

Advantages and disadvantages

Screened subnets are more secure than screened hosts, but can also allow the same "trusted" applications to bypass the application gateway. For this reason, they are listed above as the next most trustworthy to the dual-homed gateway firewall. The screened subnet is a fine place to locate information servers, modem pools, and inherently risky systems such as NIS, NFS, and X-Windows applications. These could connect to site hosts through the application gateway and inner router. Table 10.4 itemizes the benefits and drawbacks of using a screened subnet.

Modem Pools

Modem pools consist of a bank of modems connected to a communications or terminal server. Sometimes a general-purpose host system is used. Dial-in users must enter a password or use an authen-

TABLE 10.4 Advantages and Disadvantages of Screened Subnets

Advantages	Disadvantages
No site is directly reachable from the nonsecure network	Depends on routers for a large portion of the security provided
Routers send traffic to specific hosts; the application gateway need not be dual-homed	Requires a third device, the inner router
Faster throughput to inside networks via the inner router rather than a gateway	
Appropriate for high-traffic sites or those requiring high speed	
Both routers must be compromised in order to gain direct access to or from the nonsecure network	
Only the application gateway, e-mail server, and any information servers need be known to the nonsecure network	
The application gateway can support inbound one-time password electronic authentication	
Separating packet filtering from the application gateway simplifies the configuration, enhancing security	
Can permit "trusted" applications, adding flexibility	Can permit "trusted" applications, opening a potential security hole

ticating token to gain access. The user then generally TELNETs into the desired host.

Pools concentrate modems in one place, which is a good idea. This gives us a single point of control. However, realistically, users coming in from the outside with notebook computers containing PCMCIA modems who connect to a protected network represent a risk since they and their modems can bypass the firewall entirely. They must not be connected to the LAN when dialing out.

Treat dial-in modem pools as you would any nonsecure connection. This approach is shown in Fig. 10.10, illustrating a screened-host firewall with an external modem pool. This means filtering and forwarding modem traffic just as you would from the Internet. Additional inbound authentication is provided by the application gateway. Outbound, inside hosts can be prevented from using the modem pool by the packet-filtering router.

Attackers from the nonsecure network can target the modem pool itself. If they gain access, they can then use it to attack other systems. Prevent this by setting the communications or terminal server to reject connection attempts to anything other than the application gateway.

While any of the three configurations we have discussed can support modem pools, the dual-homed and screened-subnet arrangements are preferred. Figure 10.11 shows a modem pool located on the screened subnet. This allows control over the pool's use whether originated from the inside or outside. The outer filter prevents connections from anything but the application gateway. The

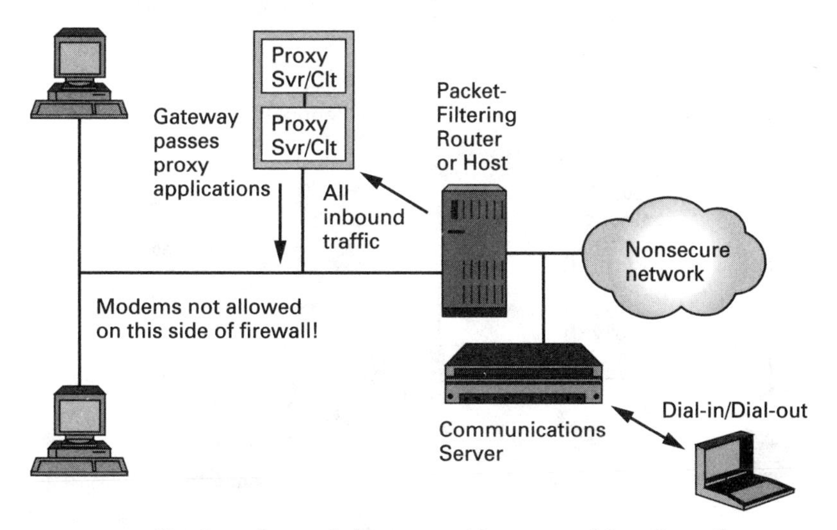

Figure 10.10 Simple modem pool placement with a screened-host firewall.

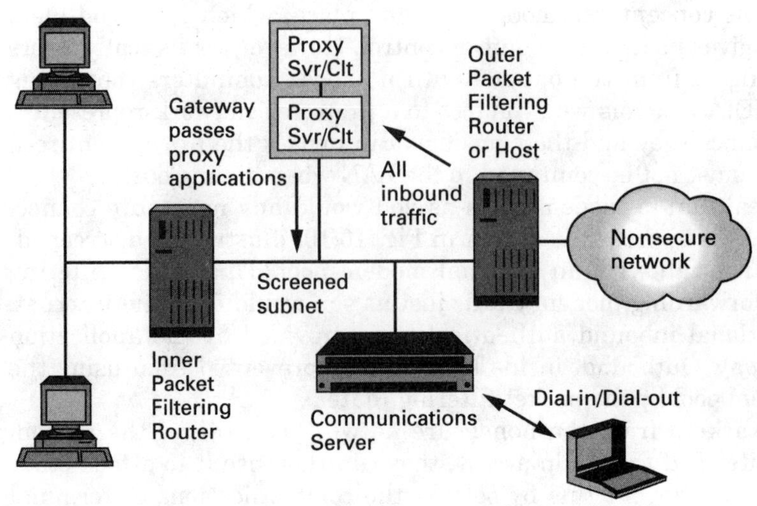

Figure 10.11 Modem pool placement with a screened subnet firewall.

inner filter forces outbound modem requests to go through the gateway as well.

In the dual-homed gateway shown in Fig. 10.12, the gateway restricts access to the modem pool from the inside. Like the screened subnet, outside access to the modem pool is prevented by the filtering router.

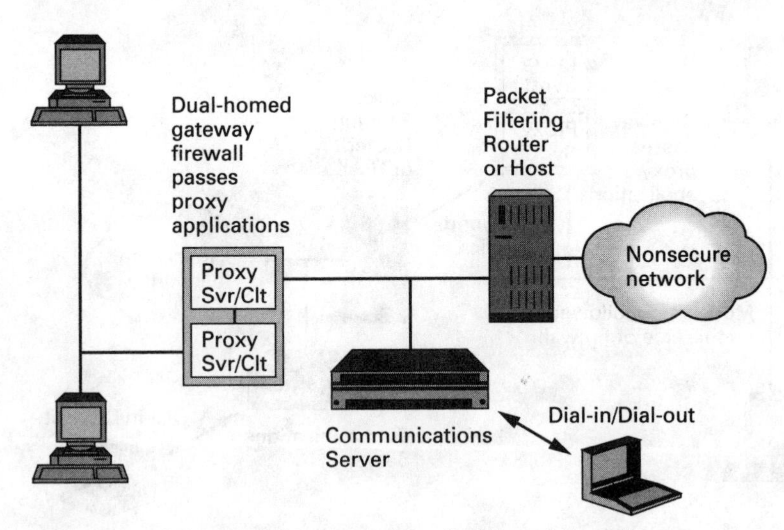

Figure 10.12 Modem pool placement with a dual-homed firewall.

In all configurations, dial-in access is restricted by advanced authentication provided by the application gateway at a minimum. Additional authentication through the communications server may be unnecessarily redundant. However, if dial-in access is allowed directly to the Internet, then this additional security is warranted.

A last reminder: Hosts equipped with modems must not be connected to the secure network. They can bypass all your firewall measures. You may be able to control the desktop machine configurations, but not the mobile user who is visiting for a day. If they choose to use their own machine, connect them to the screened subnet or let them dial in as usual. Alternatively provide them a workstation on the secure network that is not modem equipped.

Firewall Summary

Table 10.5 summarizes the choices of firewalls open to you and the characteristics of each.

Building Your Own Firewall

Most users purchase firewall equipment. Purchase costs are approximately $20,000 to $30,000 for software and training, $10,000 to $15,000 for gateway hardware, and about 12 percent of the software price per year for maintenance.

As a rule of thumb, if a purchased device will do 80 percent of what you need, then you cannot justify building your own firewall. If you cannot find a commercial device that meets your security policy requirements, then you may want to consider building your own. This is not as daunting as it might seem since there is generic and free software such as SOCKS and Karlbridge available to help you get started. We discuss these shortly.

Before deciding to build your own firewall, ask the following questions:

- Why should we build a firewall? Is there no commercial device that can meet our needs?
- Who will build the firewall?
- Who and how will it be tested?
- Who will maintain the firewall, i.e., do backups and repairs?
- Who will add new services such as proxy services?

TABLE 10.5 Firewalls at a Glance

Type of firewall	Characteristics
No firewall	■ Subjects site to chronic attacks and disruption ■ Damage may cause lost business; opportunity cost ■ Legal and fiduciary expense ■ Amounts to sponsoring attackers
Packet filters (network layer)	■ A first defense only ■ May yield too much or too little control ■ Complicated filtering tables ■ Uses up router computing resources ■ Falsification of addresses is easy ■ Works poorly with NIS, NFS, RPC, FTP ■ UDP filtering is nearly impossible
Circuit-level gateways (application layer)	■ Acts as a relay ■ Hides internal network names and addresses ■ No or some native applications ■ No application support means blocked service ■ Can add sophisticated authentication ■ Not easy to adapt to new services
Application gateways (application layer)	■ Has all the features of a circuit-level gateway ■ Runs applications natively ■ Requires changes to client software ■ Very secure ■ Can also perform packet filtering

- Can we support this additional and critical task in a timely way?
- Where does this task stand amid our other obligations?
- Who will perform user support, training, and documentation?
- What happens when this person or these people move on? How do we assure continuity of support?
- Who will watch the watchers: How do we know that they are not building in their own back door?
- Do we have the expertise to do this? Or must we rent or buy it?
- Is this a "crystal box" solution that will let us see what is going on or a less preferable "black box" solution that is easy to install but hard to customize?

Clearly, building your own firewall will be labor-intensive. By purchasing a firewall you essentially outsource this project. In fact, outside value-added-network vendors such as Advantis can provide custom firewall services without having to dedicate your internal staff. Outsourcing is an in-between option, neither fully in-house nor an off-the-shelf purchase. Firewall vendors also have support teams to customize your installation. These options should be thoroughly investigated before deciding to build your own.

Benefits

Building your own firewall may be cost-effective if you have the means to do so. You can tailor the machine to your exact needs and you will fully understand and control all aspects of the firewall. No outside vendor can know your site or your needs as well as you. You may find it advisable to engage a consultant who specializes in this area to assist you. Direct them, but let them carry the majority of the weight.

If you use public-domain software, it has been used and tested extensively. Since the source code is available, it is subject to intensive peer review. Neither statement is necessarily true of commercial products.

Drawbacks

Building a firewall is expensive, but the ongoing maintenance expense can be greater over time. The total cost to build, document, extend its functions, and maintain the firewall must be considered. Commercial products are updated more often. You can hold the vendor responsible when things go wrong. Since the source code is not made public, attackers have to work harder to find leaks.

For some homegrown implementations, the course can be grueling. In some cases, you will have to find all the appropriate daemons (we assume Unix because most public software is written for Unix). Each supported application may require special client software and may be available for DOS and Unix machines, but perhaps not for Apple System 7, OS/2, and so on. Since daemons and clients interact, as the daemons come out with new features, it may be necessary to change the clients as well.

Existing Firewall Software

There are at least four sets of software publicly available that can form the basis of your own firewall. SOCKS is an application gateway and Karlbridge is a simple packet filter. Screend and the Trusted Information Systems (TIS) firewall toolkit are public-domain software as well. We discuss SOCKS first.

SOCKS

SOCKS was originally a development name for SOCK-et-S. Somehow, the name has stuck. SOCKS is a very robust circuit-level application gateway firewall. It relays TCP segments. Application-level changes are not required. It requires modifications to a client program's TCP in order to work. SOCKS includes specific directions on how to "SOCKSify" a program. This is not very complicated and most programs can be "SOCKSified" if they use TCP.

As a TCP relayer, SOCKS does not support UDP directly, nor does it support Archie. A separate UDP relayer is publicly available (see the chapter Appendix Resource List for details). Windows applications are supported through PC SOCKS Pack or Trumpet Winsock. Otherwise unmodified X-Windows clients can use SOCKS as well. In fact, NCSA X Mosaic comes with SOCKS support built-in as an option that can be enabled. Macintosh software is also available.

The performance of SOCKS is quite good, since it stays out of the applications themselves. By working with TCP only, nearly any TCP-dependent program can be made to run under SOCKS, adding to your flexibility. Further, advanced authentication measures can be used along with SOCKS since such measures are usually invoked at the application layer. However, SOCKS does not natively support protocol-specific logging.

Karlbridge

Texas A&M University has a site (listed in the Resource List in the Appendix at the end of this chapter) that offers a variety of security tools. Karlbridge, a PC-based screening router kit, is available via FTP from Ohio State University. Site details are listed in the Appendix.

Screend

A version of Digital Equipment Corporation's Screend daemon is available for BSD/386, NetBSD, and BSDI. Screend provides packet-filtering security. It is available from gatekeeper.dec.com via FTP.

TIS Firewall Toolkit

Trusted Information Systems offers a free dual-homed Firewall Toolkit that is fairly popular. It can be downloaded from ftp.tis.com from the pub/firewalls/toolkit directory. The toolkit includes a TELNET gateway that eliminates the need for special client software. However, it does not support any special graphical user interfaces you may be using. See also TIS's WWW site listed at the end of the chapter.

TCP/IP Site Security

TCP/IP site security is an entire topic unto itself, and replicating all of it here makes little sense, especially since the threats evolve constantly. Much information is published about site security and is freely available on the Internet. In particular, obtain a copy of the *Site Security Handbook,* RFC 1244, and *Guidelines for the Secure Operation of the Internet,* RFC 1281, via anonymous FTP from any of several sources, the easiest being nic.ddn.mil in the rfc directory or from cert.org (192.88.209.5)/pub. Also look in the sec directory for security bulletin postings.

Flaws in TCP/IP

Many TCP/IP services and implementations are flawed from a security standpoint. Such flaws may be inherent to the protocol's conceptual architecture, implemented in such a way as to create a security hole or both. Either or both ways, these flaws create the potential for security breaches. For this reason, firewalls consisting of routers that filter out these services are recommended by the Computer Emergency Response Team (CERT). Many providers of Internet access such as Performance Systems International, BBN Planet, and Uunet Technologies offer such packet-filtering firewall service as an extra-cost option.

As Table 10.6 shows, CERT recommends filtering the following services due to protocol flaws and/or ongoing administrative problems.

Filter out services not needed by external users. CERT uses the example that it filters out TELNET connections when all its members are in the office. In fact, all trusted machines should be on the secure side of the firewall. If they are not, then you risk a *spoofing* attack in which an attacker pretends to be one of your outside hosts. FTP connections are also disallowed except to cert.org, which is an anonymous FTP archive. Create separate FTP and TELNET accounts for individuals and not groups as a way of retaining an audit trail.

TABLE 10.6 Service Filters and Associated Ports

Service	Port
Domain name service zone transfers*	53
Trivial file transfer protocol daemon†	69
Link (often used by intruders)	87
Sun remote procedure call	111
Network file system	2049
Berkeley Software Distribution "r" commands	512, 513, 514
lpd	515
uucpd	540
OpenWindows	2000
X-Windows	6000+‡

*Access to port 53 should be limited to known secondary domain name servers. Limiting DNS access prevents crackers from learning about systems connected to your local network.

†One cracker automated TFTP attempts (successfully) into sites that were using the TFTP daemon to boot local X terminals. Filtering port 69 would have prevented this.

‡X-Windows ports start at 6000 and extend to the highest number of terminals on the same host.

Anonymous FTP

Many sites allow file downloading to external users via this popular protocol. CERT suggests the following:

- Use the most recent FTP daemon.
- Anonymous root directories and subdirectories should not be owned by the FTP account or even be in the same group.
- Use dummy password and group files to keep tamperers out.
- Writable directories in anonymous FTP are possible and reasonably safe if CERT's recommendations are followed. These include using a modified FTP daemon, using protected directories and using a single disk drive.

For more information, use anonymous FTP to cert.org and retrieve these files:

CA-93:06.wuarchive.ftpd.vulnerability

CA-92:09.AIX.anonymous.ftp.vulnerability

CA-88:01.ftpd.hole

TCP/IP routers can limit the exchange of address information using the exterior gateway protocol. Unfortunately, the same capability does not exist for NetBIOS or Novell's IPX/SPX.

Strengthened TCP/IP Protocols

TELNET has been strengthened by CERT-IT (Italy) to provide secure client–server communication using DES, triple DES, and IDEA. For more information send an e-mail request to stel-beta-test@dsi.unimi.it and ask for more information on Stel (Stel stands for secure TELNET).

Secure FTP is now available with one-time pad encryption. Information is available at http://www.elementrix.co.il and you may download the software from ftp://ftpelementrix.co.il/pub/secftp1.zip using anonymous FTP.

For World Wide Webs, the Secure Socket Layer (SSL) provides security for all Internet application protocols including FTP and HTTP (Hypertext Transport Protocol). Secure HTTP provides security for HTTP only. Both use public-key encryption developed by RSA Data Security, Inc. You can FTP to rsa.com for further information. Actual software is being produced by Terisa Systems. The SDK (Security Data Kit) supports both SSL and S-HTTP.

Typical TCP/IP Attacks

Of the many attacks that can be launched against TCP/IP systems, a few stand out. They are:

- IP spoofing, in which an attacker pretends to be a trusted host
- Desktop or notebook-based operating systems that can act as unwitting servers
- IP tunneling, a Trojan Horse attack in which an improper transmission is buried inside a proper one
- Probing tools, including SATAN (not the virus), that can be used against your site

We discuss these as well as prevention methods below.

IP spoofing

A discussion of IP spoofing will make believers out of any remaining readers who still think that a source IP address is sufficient to authenticate a user. The technique was originally described in 1985 by Robert Morris, an AT&T Bell Laboratories researcher. You can obtain a copy of his paper as well as a related paper by Steve Bellovin via anonymous FTP to the Internet address in the Resource List in the Appendix at the end of this chapter.

As we have said, an IP address contains two parts: A source address and a destination address. A router examines the destination IP address, consults its routing table and makes a decision about which path the packet should follow. The source address is not examined during this time, leaving the way open for the next part of the spoof.

When a packet arrives at its destination, the receiving host looks at the source IP address to see who originated it. It is here that the first part of an IP spoof occurs: If an attacker substitutes another trusted IP address (determined by monitoring the target host to determine trusted relationships) for a legitimate source address, then the first part of the penetration is successful. Freeware is widely circulated on the Internet to simplify this task. Once accomplished, the receiving host now interprets the source of the packet as the attacker, not a legitimate user. The first part of the spoof is now complete.

From the above you can see once again that source IP addresses are insufficient to validate a user. Changing an IP address on a computer is easy. Thwarting such attacks requires cryptographic authentication using a one-time randomly selected password to truly validate a user. The TCP header and message following the IP header must be encrypted to avoid giving visibility to TCP sequence numbers and the message itself. Some systems even encrypt the entire IP packet and then replicate the IP header in plain text so the packet can be routed. These techniques will be discussed in greater detail below.

The second part of the spoof is harder. In TCP, sequential acknowledgment numbers are used in each positive acknowledgment to assure the sender that TCP segments have been received properly. If the spoofer can predict the sequence numbers that would be used between two legitimate hosts as they establish a connection (and some Unix machines have very predictable numbering routines, making the task much easier), then the spoofer can send packets to either of those hosts by first forging the source address of the other machine as well as sending along predicted sequence numbers. Then the other host will think it is receiving from its legitimate partner when in fact the spoofer is at the other end.

On the outbound side, the spoofed machine will still be sending to the legitimate user, whose destination address has not changed. This means that the spoofer can send commands to the fooled machine, but cannot see anything coming back. However, actions with predictable responses, such as attempts to change access control files, reduce this problem to the level of an inconvenience rather than a showstopper.

Once completed, access to other machines inside your network is often possible. The attacker can roam your machines at will. Unfortunately, IP spoofing attacks are almost impossible to detect.

Prevention

One simple technique is to set your routers so that they will not accept an internal IP address from outside your network. Loose or strict-source routed packets must not be accepted by the router from an external network. Trusted relationships, such as those established by the popular Network File System, cannot be permitted between internal and external machines.

Outbound, filter out source addresses that are not internal IP addresses. More recent Bay Networks router software supports this feature, as does Cisco Systems software. Among others, both Livingston, Inc. firewalls and Cabletron Systems LAN Secure firewalls support this feature.

Some routers can search out discrepancies between IP addresses and the network from which the message originated. One such product is CyberGuard FireWall from Harris Computer Systems.

The above discussion applies to the current version of IP, version 4. IP version 6 will simplify some of these security problems but not mitigate them completely.

Anti-IP address spoofing techniques

Cisco Systems offers transparently operating software that can defeat IP address spoofers in two ways. The first method simply encrypts the content of IP packets but not the IP header. This makes sequence numbers invisible, but does not prevent IP address spoofing. It is a partial solution and may well be sufficient, but is not totally secure from spoofing attack.

The second technique takes the entire IP packet, encrypts it, and then places the packet's IP header at the front, as shown in Fig. 10.13. At the receiver, the plaintext IP header is set aside, the message decrypted, and the now-plaintext IP source address is used. The decrypted source IP address can be compared to the preceding plaintext source address to see if they match. If they do not, this signals the start of an IP address spoofing attack. The negative aspect of this approach is that the IP header is sent twice, increasing overhead.

Encryption is performed using the two-key public-key system employing the well-known Diffie–Hellman key-exchange technique described earlier in Chaps. 4 and 5. Hashing of the data, plus a

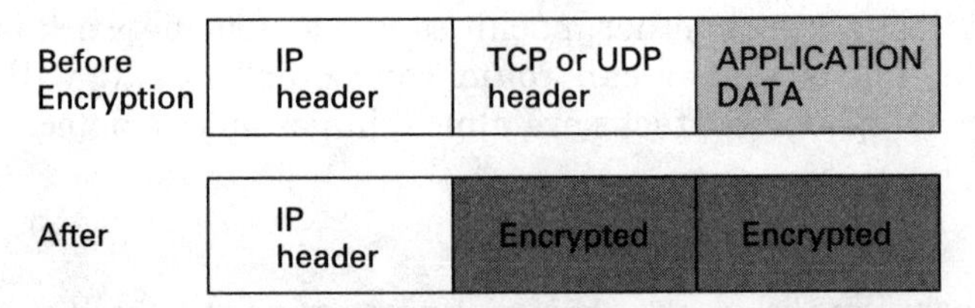

The IP header remains in plain text so the packet can be routed.
This does not protect the source or destination address against a spoofing attack.

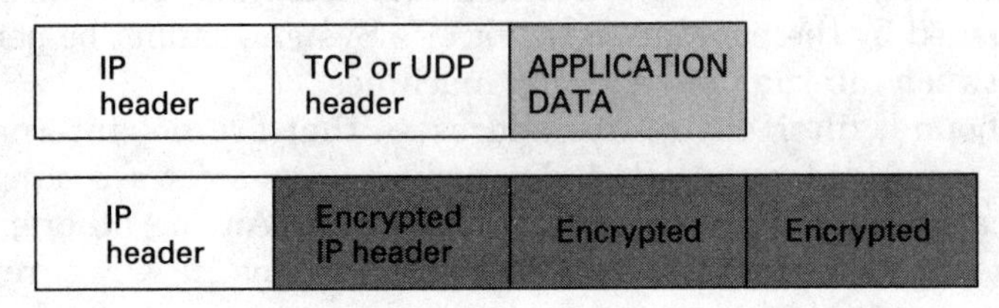

Here the IP header is encrypted, preventing a spoofing attack.
The decrypted header can be compared to the plain text header to detect such an attack.
Note the additional overhead incurred.

Figure 10.13 Header encrypting methods.

digital signature using the federal government's Digital Signature Standard, ensures that the message was not tampered with and that the source of the message is legitimate. Encryption is performed using the federal Data Encryption Standard.

In order to manage the system, a separate management station is needed to distribute electronic keys to the routers involved. It also collects statistics on usage and allows diagnostic insight into the system's functions. Public-key certificates are issued by the machine that specify security clearance levels as established by the system's owners.

The above software is built into a Cisco router but a similar solution is also offered by other vendors. A more integrated solution requires the development of a multiprotocol router designed from scratch to route, filter, and encrypt–decrypt routable packets. Such a system, called the Security Router, is offered by Network Systems of Minneapolis, Minn. Security Router is a midrange router that routes protocols including IP, Novell Internetwork Packet Exchange, Apple's Datagram Delivery Protocol, DECnet Phase IV, and Xerox Networking Services. It provides internal encryption using a variety of algorithms including the Data Encryption Standard, the International Data Encryption Algorithm, and NSI's proprietary algorithm, Network Systems

Cipher 1. Encryption keys are verified using the RSA public-key system. Since a routable header is attached to the front of each packet, traffic passing through the machine is routable by any other IP-equipped router.

The device can also filter packets at the network layer and act as a frame-filtering bridge at the OSI data link layer. The Security Router is therefore a bridge–router or "brouter." If you cannot use this feature, then you may end up paying for something you do not need.

Two attributes make this integrated approach unique. The first is that the device must have the processing power to do all these high-overhead operations at LAN and high WAN speeds such as T-1 at 1.544 Mbits/s. The second aspect is economic. An integrated solution such as this one costs around $12,000 to $18,000 depending on how it is equipped. When one considers that an outboard encryption device alone costs around $10,000 per end and must pass plaintext through a router, thereby opening a security hole, the integrated approach takes on greater appeal from a security and flexibility aspect as well as from an economic standpoint.

This is especially true in a mesh network where the number of point-to-point encryption devices greatly outnumbers the quantity of routers used. For example, a triangle of routers would require three routers at a total cost of perhaps $45,000. But six encryption devices would cost $60,000 and not give as much security or flexibility, as in being able to route securely through the Internet to another Security Router in a remote office.

TCP/IP implementation-dependent security features

Some vendors implement additional security features for their TCP/IP protocol products. One such mechanism is offered by TGV of Santa Cruz, Calif. TGV's MultiNet for Windows TCP/IP software, Secure/IP, works with Windows-based client workstations. The basic mechanism is a one-time password issued by a hardware "token" card such as Bellcore's S/Key, Digital Pathways' SecureNet, or Security Dynamics' SecurID.

When a user in this environment attempts to log in, Secure/IP asks for a user's personal identification code. If the code is correct, Secure/IP then asks for a one-time password from the token (the card) and the login proceeds.

While MultiNet already supports Kerberos authentication, described elsewhere in this book, Kerberos does not scale well into wide area networks. The reason is that Kerberos requires a central repository of encryption keys, which becomes cumbersome in a WAN, considering the number of keys and who will administer and maintain them.

Secure/IP is aimed at WAN-based applications, including dial-in users and Internet connections. Secure/IP runs on DEC VAX and Alpha processors.

Operating system weaknesses

Any user who opens a TCP/IP connection directly to the Internet creates the opportunity for an outsider to use that connection to gain access to the internal network. This is especially true for Windows 95 users and OS/2 3.0 Warp users since both operating systems have built-in Internet gateways. All it takes is clicking on the icon. In fact, the flaw holds for Windows for Workgroups users and any multiuser environment that uses the operating system SHARE command as a default. Any network server or peer-to-peer network connected to the Internet can also act as an unwitting gateway unless a firewall is installed between it and the Internet. SHARE offers no security features at all.

Documentation in Windows 95 warns users of this potential hazard. As for deterrents, the best solution is to tie Internet access into the security offered by your network operating system. This means that all would-be Internet users must first pass through the security screening on a central server. For example, the NOS could compare access names with those in a Novell NetWare bindery. A second solution, discussed earlier in this chapter, is to use an application gateway to limit strictly the applications that can be used between inside and outside. Other filtering devices such as a packet-filtering firewall or even a bridge with filtering capability (however limited) are better than nothing. FTP is a special hazard because it can be used to traverse the Internet and gain access to systems that are not protected. Without safeguards, it is possible for users to enter the Internet and then come back into an area where they should have no privileges.

As a host on the Internet, your machine is accessible by others, especially if you are dialing in through a modem. Unless disabled, others can "ping" your machine (to see if you are on the network), TELNET into it, and utilize FTP if your machine is so enabled. The moral could not be simpler: Restrict or turn off these functions if you want to keep outsiders *out*.

Tunneling

Today, it is common to tunnel one protocol inside another. This can give an attacker an opportunity to create a Trojan Horse attack by secreting their code inside yours. For example, it is possible to place an IP packet as a message part in a Domain Name Service or TELNET message. To prevent this, avoid giving an attacker an entry point: Encrypt between nodes. Tunnels can originate from a person in your organization or in another.

Software Tools: SATAN

In 1995 two security experts, Dan Farmer and Wietse Venema, wrote a software tool that probes for network security weak spots. Farmer placed this code, titled SATAN for Security Administrator Tool for Analyzing Networks, on the Internet (available via anonymous FTP to ftp.mcs.anl.gov in the /pub/security directory) where it is freely available to everyone, good and bad.

Farmer also authored and made public COPS, a software utility standing for Computer Oracle and Password System that scans for password security weaknesses. Neither COPS nor SATAN is by any means the only hacking tool available, but SATAN is the most complete and comprehensive of its type. Do not confuse this SATAN for the virus Satan of the same name.

SATAN makes use of known security tricks in Unix and generates well-organized reports about what it finds. Its authors wrote it in the belief that SATAN will encourage site managers to give network security higher priority since attackers are free to use it against them. Site managers are encouraged to use it themselves as a software version of a tiger-team attack. What Farmer and Venema hope to achieve is to make security problems easy to find and fix. Giving SATAN to attackers as well supplies motivation. Apparently their view is that while their software will make some break-ins possible, it will sensitize more network managers and on balance prevent more break-ins than its use will allow.

Since SATAN is available on the Internet, moral arguments about the propriety of releasing such code are moot. It is there and you can make it work for you because it may well be used against you.

SATAN in detail

SATAN uses a MOSAIC-like graphical user interface, not unlike a World Wide Web browser. Unlike other similar tools on the

Internet, SATAN is easy to operate but does require some Unix knowledge as well as a copy of Perl5, a scripting language. Searching by domain name, SATAN can scan any specified network. Once it finds a single site with a security hole, it can then probe through it, even if hundreds or thousands of TCP/IP capable machines reside behind it. SATAN then creates a report that divides the machines scanned into classes of vulnerability. Servers that run TCP/IP, including LANs capable of Internet connections, are vulnerable to outside probes even if the server is based on a non-Unix operating system such as NetWare or Windows NT. SATAN identifies all known vulnerabilities about the operating system in each machine. It then probes for those vulnerabilities and reports the results, machine by machine. SATAN even advises the operator how to close the holes found in each machine.

SATAN tries to start connections with target computers by trying remote port numbers that equate to TCP/IP applications such as DNS, TELNET, FTP, NFS, and TFTP. It also seeks out weaknesses in password files and rlogin-enabled systems.

Some network security specialists run SATAN fairly often as a quality check, letting it tell them of security holes so they can be plugged. Unfortunately, SATAN also points out weak spots for which there are few or no defenses. It automates the otherwise tedious and error-prone task of having to test manually and redundantly. The result is heightened awareness, corked leaks, and an understanding of how an attacker might use SATAN against their network. Seen in this light, security managers could be considered negligent if they fail to test their networks periodically with SATAN. But conscientious administrators will have little reason for fear.

There have been some reports on the Usenet security forum that SATAN can crash certain Sun Microsystems workstations by starting daemons (processes) that then go into infinite loops. One such reported daemon is rpc.ttdbserverd, which is a Sun ToolTalk database server. Many system managers regard this as the result of an unstable process rather than an attempt by SATAN to crash the machine.

Point your World Wide Web browser at http://csrc.ncsl.nist.gov/first/satan/satan.html to find out the latest information. You can also get information via FTP to ftp.cnidr.org. Think-tank SRI Inc. of Menlo Park, Calif. offers a report analyzing SATAN's capabilities through its International Information Integrity Institute.

Defeating SATAN with COURTNEY

For every measure there is a countermeasure. COURTNEY is a software tool that detects the use of SATAN against a site and triggers alarms. The alarms are tripped by the repetitive actions of SATAN as it figuratively knocks on your door in different ways over and over. COURTNEY and other anti-SATAN tools, some of which can trace a SATAN probe back to its source, are available at coast.cs.purdue.edu in the /pub/tools/unix/satan/defenses directory. It seems prudent to run COURTNEY as a background task, as another set of eyes on your network.

Defeating SATAN with firewalls

Since SATAN probes by port number, packet-filtering firewalls can screen out probes to undesired ports.

Pingware

SATAN is not the only network-probing tool available. Bell Communications Research sells a security tool called Pingware. While Pingware is similar in function to SATAN, Bellcore interviews potential licensees first to ensure that the software will not be used by attackers.

TCP/IP Wrappers

Individual host machines can be protected by installing software that authenticates the incoming packet before executing its contents. Thus the request is wrapped in an authenticator. Variations on this theme include encrypting the packet either including its header and replicating it in plaintext for comparison after decryption or simply passing the packet header in plaintext and encrypting upper-layer information only.

Incursions

If you should experience an attack, whether successful or not, notify the Forum of Incident Response and Security Teams (FIRST). There is an incident form posted at cert.org and first.org. These sites can be reached via Gopher or FTP (login as user "anonymous"), and first.org has an http server. Use http://www.first.org/first for access. You can obtain assistance and advice from these sites, and can become a member of FIRST as well.

Conclusion

Since most of this chapter has been technical, a few closing notes on pertinent nontechnical issues seem appropriate.

Build a policy first

The most important task is to *build a policy* from which a firewall strategy can be derived. The second task is to decide on the necessary auditing and logging needed to satisfy the auditors. The third task is to build the firewall at a reasonable cost.

Firewall costs

Nothing is free. Even "free" software carries an installation cost in terms of staff time that may range up to $30,000 or even more. Commercial software is even more expensive: DEC's SEAL starts at $25,000 and very-high-end systems cost $100,000. Since cost is a direct reflection of policy, the policy may be modified several times in order to bring the cost and realistic security level into balance.

Administrative costs

Effective firewalls need constant tending, such as checking activity logs for attempted attacks. Therefore effective firewall administration is essential to having an effective firewall. Remember that the threat constantly evolves upward.

Having a firewall is no excuse for laying back on system administration. Indeed, solid system administration is your backup in case the firewall is breached. Administration includes periodic testing of the firewall using SATAN or equivalent and having a procedure in place to deal with a breach should one occur.

Bypassing the wall

A firewall will not protect you against a virus or theft via diskette ("sneakernet") or modem that bypasses the firewall. Nor will a firewall protect against stolen or compromised passwords. Passwords are sent in *plaintext* across the Internet. The level of sophistication is such today that passwords (on WANs at least) are becoming obsolete and security tokens using one-time passwords are needed to provide valid authentication.

Firewalls can do nothing to stop data-driven Trojan Horse attacks such as tunneling or burying an executable program in

Sendmail or the like. Similarly, MBONE (Multicast IP transmissions) can contain Trojan Horses. WWW, Gopher, and WAIS can also be used as bases for data-driven attacks. As yet, the detailed means to secure these risky protocols are not completely clear, but placing such servers on a screened subnet accessible through an application gateway seems adequate.

Firewalls cannot replace users who are not security-conscious. As noted elsewhere in this book, users have a responsibility, too.

The last resort

If you have top-secret information, do not connect to the Internet at all!

Review

- Firewall devices include intermediate networks, packet filters, circuit-level gateways, and application gateways.
- One prime purpose of a gateway is to keep vital data inside the organization.
- Another prime purpose is to limit legitimate users' access to external and internal databases and applications.
- Network administrators use firewalls as a way to implement a security policy. They give that policy force.
- Firewalls are advantageous in that they are a single point of contact between the Internet and the interior network. This makes the network much easier to manage effectively.
- Threats always evolve. Even the strongest firewall countermeasure will be subverted over time. Countermeasures must therefore evolve, too.
- Risk management recognizes that theoretically possible attacks can go well beyond what is practical. By making the work factor high enough, we push practical attacks into the theoretical category, making the potential gain smaller than the sum of the expense, risk, and effort required. Breaking past an application gateway is one such example.
- The expense of avoiding all risks is prohibitively high and is unnecessary if proper risk management is followed.
- Security is a systems problem, meaning that a combination of people and machines creates security risks as well as thwarts them. A top-quality firewall can be compromised through sloppy administration.

Appendix 1. Resource List

Note: Site names and/or directory paths appearing at the end of a sentence may be enclosed in double quotes to specifically exclude the period at the end of the sentence. Do not use the double quotes as part of a name or path! Where appropriate the period at the end of the sentence is simply omitted to avoid mistaking it for part of the site or path string.

General

Get the Internet firewalls frequently asked questions (FAQ) from ftp.greatcircle.com in pub/firewalls.

You may subscribe to the Internet firewalls mailing list by sending the message "subscribe firewalls" in the body of a message (not on the "Subject:" line) to "Majordomo@GreatCircle.COM". This site also has firewall mailing list archives located in pub/firewalls/archive.

There is an Internet firewall toolkit at ftp.tis.com in pub/firewalls. Papers published on the subject are also listed. For papers on firewalls and break-ins, see research.att.com in dist/internet_security

RFC 1108 (*Security Options for the Internet Protocol*), RFC 1244 (*Site Security Handbook*), and RFC 1281 (*A Guideline for the Secure Operation of the Internet*) are available from ds.internic.net in rfc/ ; nis.nsf.net in the internet/documents/rfc file or at "nisc.jvnc.net". The form of the rfc listing is rfc/RFCnnnn.txt.version

Look in Usenet under "comp.security.misc". Read especially the security.faq (frequently asked questions). Also see alt.security.

The CERT FTP site at cert.org has an enormous amount of useful, specific security information. Start with the security.faq here as well. You can join the CERT mailing list by sending e-mail to "cert@cert.sei.cmu.edu". Advisories may be retrieved from the same e-mail address in /pub/cert-advisories via anonymous FTP.

CERT advisories are offered through at least three sources:

Internet mailing list: cert-advisory-request@cert.org

Anonymous ftp: cert.org (192.88.209.5)/pub/cert-advisories

Usenet: comp.security.announce

This site has several mirrors at corsa.ucr.edu, cert.sei.cmu.edu, info.cert.org, and ftp.cert.org.

You may subscribe to the Defense Data Network Security Bulletins by sending a mail request to "nic@nic.ddn.mil". Past bulletins are posted at nic.ddn.mil in the sec directory.

Subscribe to the Risks Forum via mail request to risks-request@csl.sri.com.

Use FTP to access the RSA Data Security server at "rsa.com". Assorted documents relating to the Clipper chip, frequently asked questions about public-key encryption, and related RFCs are stored on this machine. Public-key security mechanisms will likely become the means to assure the validity of money transactions via the Internet. Certifying the validity of such keys could be a $50 billion business by the year 2005 and has attracted the attention of major banks and credit card firms.

General information regarding security is available at "ftp.gsfc.nasa.gov". Another government FTP site is csrc.ncsl.nist.gov.

For information on viruses, send a message to "listserv@lehigh.edu". The message content should read: SUB VIRUS-L (your name).

SATAN

There is very good security information available on "coast.cs.purdue.edu", especially on SATAN and anti-SATAN tools. Another source is via FTP to ftp.cnidr.org.

Firewalls

By all means get a copy of *Keeping Your Site Comfortably Secure: An Introduction to Internet Firewalls,* written by John P. Wack and Lisa J. Carnahan. This very thorough document is available from the National Institute of Standards and Technology as NIST Special Publication 800-10. You can view, copy, or print it from their WWW server at http://csrc.ncsl.nist.gov/nistpubs/800-10

Obtain a copy of *Firewalls and Internet Security: Repelling the Wily Hacker* by William R. Cheswick and Steven M. Bellovin, published by Addison-Wesley, Reading, Mass., in 1994. The authors are researchers at AT&T Bell Laboratories and are well-known experts in the field.

The University of Texas at ftp.sesqui.net offers RFCs and information on firewalls.

Texas A&M offers security tools including screening router software at FTP site net.tamu.edu. Look in pub/security/TAMU

An FTP site with a great deal of firewall information is ftp.greatcircle.com; alias mycroft.greatcircle.com. Look in the /pub/firewalls.digest.

Try ftp.delmarva.com for security information in Postscript form; Raptor Systems firewall systems documents and mailing list, and assorted security documents.

SOCKS

See ftp.nec.com in pub/security/socks.cstc for information on SOCKS. Use file://ftp.nec.com pub/security/socks.cstc if working through MOSAIC. Look in the file labeled FILES for a description of the directory's contents. Also look in ftp.butler.hpl.hpcom for more information. You can obtain a copy of SOCKS via anonymous FTP from ftp://ftp.nec.com/pub/security/socks.cstc/socks.cstc.4.2beta.

Basic clients for TELNET, FTP, FINGER, and WHOIS are included. Be sure to get the FAQ as it includes locations of other related software, the mailing list location, and other useful information.

Karlbridge

Karlbridge is available via FTP from nisca.acs.ohio-state.edu in the pub/kbridge directory.

IP Spoofing

For information on IP spoofing, connect via FTP to "ftp.research.att.com". Switch to the /dist/internet_security directory. Retrieve Steve Bellovin's paper, "Security Problems in the TCP/IP Protocol Suite" as "ipext.ps.z". To retrieve Robert Morris' paper, "A Weakness in the 4.2BSD Unix TCP/IP Software," retrieve "117.ps.Z". These files are both in Postscript form (denoted by the ps) and are compressed using Unix data compression (denoted by the Z).

Appendix 2. Vendor List

Firewall vendors

ANS CO + RE Systems, Inc. — Interlock
1875 Campus Commons Drive
Reston, VA 22091
Tel: (800) 456-8267 or (703) 758-8700
Fax: (703) 758-7717
interlock@ans.net

Vendor	Product
Cisco Systems, Inc. 170 West Tasman Drive San Jose, CA 95134 Tel: (800) 553-6387 or (408) 526-4000 Fax: (408) 526-4100 http://www.cisco.com	
Digital Equipment Corp. 550 King Street, LKG1-3/J17 Littleton, MA 01460 Tel: (800) 344-4825 or (508) 486-6963 Fax: (508) 486-6311	SEAL Gateway firewall
Livingston Enterprises, Inc. 6920 Koll Center Parkway #220 Pleasanton, CA 94566 Tel: (800) 458-9966 or (510) 426-0770 Fax: (510) 426-8951 Internet: support@livingston.com	Manufacturer of FireWall IRX router
Morningstar Technologies Columbus, OH Additional information is available on the World Wide Web at http://www.morningstar.com	Morningstar Express packet-filtering routers
NetPartners, Inc. 4 Park Newport Newport Beach, CA 92660 Tel: (800) 723-1166 or (714) 252-5493 Fax: (714) 644-0577 sales@netpart.com	
Norman Data Defense Systems 3028 Javier Road, Suite 201 Fairfax, VA 22031 Tel: (703) 573-8802 Fax: (703) 573-3919 BBS: (703) 573-8990	The Norman Firewall allows network entry and exit and checks packets for more than 6,500 known viruses.
Raptor Systems, Inc. 69 Hickory Drive Waltham, MA 02154 Tel: (800) 932-4536 or (617) 487-7700 Fax: (617) 487-6755 support@raptor.com	Manufacturer of Eagle Network Security Management System
Secure Computing Corporation 2675 Long Lake Road Roseville, MN 55113 Tel: (612) 628-2700 or (800) 692-5625 Fax: (612) 628-2701	Manufacturer of high-level application-level gateway firewalls

Sun Microsystems Tel: (415) 786-8199 Web sites: http://www.sun.com or http://java.sun.com	Netra Servers and SunScreen S-100 firewall
Trusted Information Systems 3060 Washington Road Glenwood, MD 21738 Tel: (301) 854-6889 Fax: (301) 854-5363 Copy from anonymous FTP to ftp.tis.com in pub/firewalls/toolkit. More information is available on the World Wide Web at http://www.tis.com.	TIS Gauntlet bastion host hardware/software firewall toolkit; permits construction of do-it-yourself firewalls.
Uunet Technologies, Inc. Falls Church, VA (703) 206-5600	LAN Guardian IP Encryptor; LAN Guardian can work with or in place of a firewall or router.

Software-based firewalls and auditing tools

CheckPoint Software Technologies Ltd. U.S. Distributor: Internet Security Corp. One Militia Drive Lexington, MA 02173 (617) 863-6400 e-mail: info@security.com	FireWall-1
Open Computing Security Group, Inc.	CyberSAFE Challenger
Preferred Systems, Inc. 250 Captain Thomas Blvd. West Haven, CT 06516 Tel: (203) 937-3000; (800) 222-7638 Fax: (203) 937-3032	AuditWare multiserver auditing tool for NetWare

Encryption devices

Cylink Corp. 310 North Mary Avenue Sunnyvale, CA 94086 Tel: (408) 735-5800 or (800) 533-3958 Fax: (408) 735-6643	
Isolation Systems 26 Six Point Road Etobicoke, Ontario, Canada M8Z 2W9 Tel: (416) 231-8561/1248 or (800)387-8706 Fax: (416) 231-8561	Hardware-based packet-encryption products using assorted algorithms. Supports frame relay and Ethernet-capable Cisco routers.

Chapter

11

Sources of Leaks in Voice Systems

"[Toll fraud] is the only crime I know where the victims are held liable."
DONALD DELANEY
New York State Police

Objectives

At the end of this chapter you will be able to:

- Distinguish data hackers' motives from those of phone hackers
- Define what is meant by toll fraud
- List the three kinds of toll fraud
- Evaluate the magnitude of the toll-fraud threat
- Describe the ways in which toll fraud can take place
- Identify hacking attack patterns and the point where hacking becomes abuse
- List possible points of entry
- Itemize proactive defense tactics
- Discuss the advantages and disadvantages of "going dark"
- Discuss FCC and congressional action (or lack thereof)
- Compare anti-toll-fraud products from representative vendors
- Formulate a strategy and tools that suit your needs and budget

- Obtain further assistance from appropriate user groups
- Discuss carrier toll-fraud prevention programs and their elements
- Outline employee abuse patterns and ways to dampen them

Introduction

Data hackers, on whom we have concentrated thus far, *steal information* and *damage systems.* Voice hackers *steal telephone service.* In this chapter we discuss the latter crime, called *toll fraud.*

Placing a value on toll fraud is difficult. There are no monthly reports by private branch exchange (PBX) owners to some government agency confessing that they have been taken. At the same time, figures produced by telecommunications consultants and vendors are self-serving. They want to make the figures as high and as alarming as possible. It is good for business. That said, the toll-fraud problem is very real, as Fig. 11.1 illustrates.

The best figures come from market research firms and industry professional associations. The International Communications Association, a respected user organization, places the annual loss at between $500 million and $6 billion. Other estimates suggest between $1 and $4 billion a year, giving us some idea of the magnitude of the problem. The loss is multiplied because stolen service is used to subsidize other thievery such as computer hacking. Industry involuntarily underwrites hacking, drug running, and other illicit activities in which stolen telephone service is used. At the beginning of 1993 an estimated 700 legal disputes with telephone companies valued at $70,000 each were outstanding—$49 million in disputed dollars.

This is not just a big-company problem. In 1992, about 60 percent of the attacks on AT&T's customers were big companies. Smaller companies are now the target, in part because big companies have locked down their systems.

Since 1992, AT&T's NetPROTECT Basic Service has forcibly removed over 2,500 hackers from their network. Though the average toll-fraud loss has decreased by over 75 percent, AT&T continues to call about 1,000 customers a month regarding suspicious calling patterns.

Data people cannot pretend this is a voice problem. Voice hackers get to data networks through the PBX, or they attack modem pools directly. *Network security is a joint data and voice responsibility.*

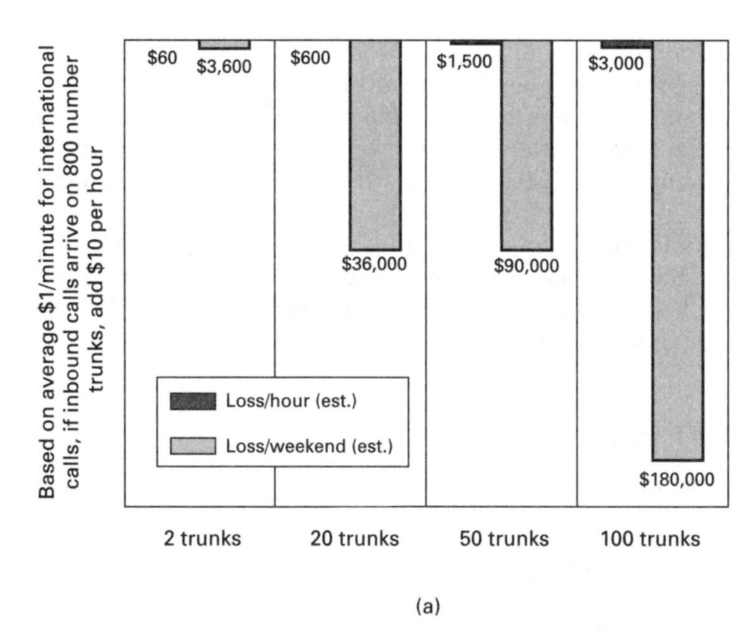

(a)

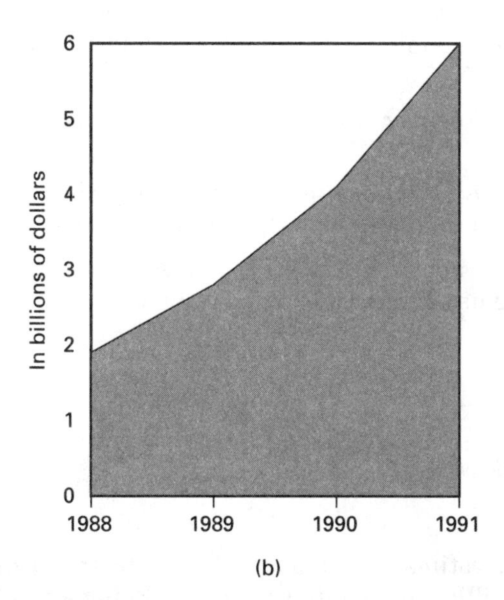

(b)

Figure 11.1 Toll-fraud losses by incident and year. (*a*) While number of trunks doesn't greatly alter dollar losses per hour, it significantly increases dollar losses per weekend. (*b*) Phone hacking caused companies to lose about $6 billion in 1991—an all-time high. (*Source: Meta Group.*)

Imagine that you come to work on Monday morning and discover the modem pool frozen, showing a continuous 48-hour connection. Perhaps the incident was an attempted data hacking effort confounded by the LAN communications server's security or perhaps it was just a hung trunk. Most people would just clear the problem and move on.

Two weeks later, your telecommunications manager gets her phone bill. That is when you jointly discover that the seemingly harmless call was a successful voice hacking effort that piled up some $40,000 in toll charges. It can and does happen.

Kinds of Toll Fraud

Toll fraud includes six kinds of abuse, in descending order of frequency:

- *PBX fraud,* where hackers assault a PBX.
- *Cellular carrier switch fraud.*
- *Calling card fraud,* where people steal credit card numbers. On the street, a stolen card number is sold for about $12 per call.
- *Local exchange carrier (LEC) switch fraud,* where people attack central offices.
- *False service subscriptions,* made from a pay phone or subscriptions made and used for a month with no intention to pay.
- Other forms of abuse, including *employee abuse,* where employees abuse the use of their telephone privileges.

PBX Fraud

"...Given the persistence of hackers, the most secure PBX is one without phones."

WOODY RITCHEY
AT&T

One consultant estimates the odds at 1 in 18 that someone will attempt to gain illicit entry into your PBX.[1] Being a small business is no protection. If you have a phone system through which users can dial out from a remote location, you are vulnerable.

What motivates PBX fraud hackers?

Hackers break into PBXs less for challenge and more for gain. They use your PBX to hack other PBXs and computer networks. Sometimes they use stolen service to call hacker bulletin boards. Some hackers steal your service and resell it: Drug traffickers are their primary customers. These evil entrepreneurs use *your* phone lines to call their contacts, many of whom are in Latin America. Hackers even sell service to other hackers.

The carriers have identified some 30 international locales likely to be hacker destinations. They watch outbound international direct-dial traffic going there plus their 800 service domestic networks for possible toll abuse. They do this continuously. In case of suspicious activity, you as the subscriber are notified.

How PBX Fraud Takes Place

Many PBXs have inbound 800 numbers for employee use. Employees use them to retrieve their voice mail and to speak to people in the office.

Another reason for having inbound 800 service is to allow access to a service called direct inward system access (DISA). DISA is used to allow an inward-calling person access to an outbound line. This allows calls to be billed at a much lower rate than if the employee used a credit card—sometimes 40 percent less. DISA can provide big savings to companies whose credit card use is high, but so many users are uncomfortable with DISA's security that they have disabled it—costing them as a whole some $200 million a year.

Hackers call 800 numbers knowing that some of them are used for DISA port access. That objective met, the new goal becomes finding the password that will grant them (hopefully unrestricted) outbound service.

Another way to gain access to the PBX is through its dial-in maintenance port. This port is used by the organization providing service for the PBX. In case of a switch problem, hackers can dial in and either fix or diagnose it. They can also do remote switch programming. Having a maintenance port is a good idea. It is useful and allows quick response to a problem. In the process, such maintenance ports create a security hole through which hackers seek to crawl. There are three basic defenses: disable the remote maintenance port (an overreaction), use a solid password (always a good idea), and use a nonstandard modem not for sale to the general public (many vendors now do this). There are also security authenticating devices sold for such ports as an additional check (e.g, from LeeMah Data Security, Security Dynamics' SecureID product, and AT&T's DataLock).

DISA and maintenance ports both require password authentication. Here the default password problem rears its ugly head again. Systems that arrive from the factory should have all passwords changed before the switch enters service. Passwords such as TEST and GUEST should be expunged. Nonsequential (e.g, not 1-2-3-4), six-character phonetic passwords should be used. Sometimes several levels of password should be employed.

Among PBX vendors the trend is to deliver switches without any default passwords. For example, AT&T's Definity G3 switch comes without passwords. The software asks the user to set passwords before they place the switch on line. Presumably owners will be creative enough to use something original, such as WOMBATKEY.

However, an AT&T report in 1994 said, "Statistical sampling indicates there is a high likelihood clients still have one or more default passwords in place on their System 75 and/or Definity System Generic 1 or 3." Whether you have an AT&T switch or not, *go change those passwords*.

How it starts

Here is the password problem that the hacker faces: Dial 800 numbers until a PBX is found. This is not a small effort as there are about 7.0 million 800 numbers in use (out of a possible 8 million). Try DISA or maintenance passwords until one works. Uniquely, *you* can make this easy or difficult for the attacker. Having found the password, start making those international calls, especially over weekends.

Clues

In the hacking stage, you will see an abnormally large number of uncompleted inbound 800 calls to remote access facilities such as voice mail and DISA ports as hackers try to crack voice mail and/or DISA passwords. The calls will be short as they try password after password. If they succeed, an increase in the number of outbound 700, 900, and 950 calls may appear in audit trail reports. Calls at odd hours will increase; typically, these had been low-activity periods.

Sometimes employees buttonhole the telecommunications manager and relate stories of strange or even obscene calls. Receptionists and operators in particular may sense something is wrong. This is a near-absolute indicator that your system is under attack via social engineering at the least.

Call-accounting systems may indicate calls originating from nonexistent "phantom" extensions or trunks. This implies that the hacker has found an unused extension. Even worse the hacker may have programmed the extension via PBX system programming. Unless the maintenance port is also monitored via call accounting, you may not know if it too is under attack.

Once a hacker gains entry, the volume and duration of domestic toll and international calls will increase markedly, especially during weekends and at odd hours.

Adding insult to injury, as hackers use more of your remote access facilities, your own people will become increasingly blocked from using your own paid-for DISA ports and voice mail.

Area Codes

Area code 809 (the Caribbean), South America, the Middle East, and Europe are the most popular hacker call destinations. Many hackers are smart. They know the symptoms their meddling can cause. They also know about call-accounting systems. If they are smart, they find a way to disable call accounting so they can carry out their mischief unnoticed. Preventing such disablement means performing call accounting separately, in a separate physical computer or a separate logical entity within the PBX.

Points of entry

These are some of the vulnerable points in a PBX:

- 800 numbers
- Voice mail
- Automated attendants
- Modem pools
- Electronic mail
- Teleconferencing bridges
- Automatic call distributors
- Remote maintenance ports

Defensive Tactics

Much as we would like, we cannot offensively go out, find hackers, and throw them in the slammer. Hackers have several undeniable advantages:

- They have the offensive. Hackers can choose their ground and the terms of the engagement.
- Time works in their favor. They have no limit on the time they can invest or when they choose to invest it. We have to be continuously on guard, a far more difficult and draining task.
- Persistence works in their favor. They can afford to be patiently persistent, and persistence is the most powerful tool there is. We, however, can be just as determined as they.
- Hackers have something clear to gain. The best we can do is retain what we already have.
- They are intelligent. Expect them to be creative and original. However, we can be just as intelligent and we have the advantage of control of our systems—if we choose to exercise that control.

Defensive tactics do not imply that we cannot be proactive. We can. The simple concept is to control everything over which you have control. Here are some tactics to consider (see Fig. 11.2).

1. *Turn off DISA ports during off-hours.* Weekends are prime hacking time. If your organization is dormant during these hours, turn off DISA access. Employees who must make toll calls can use

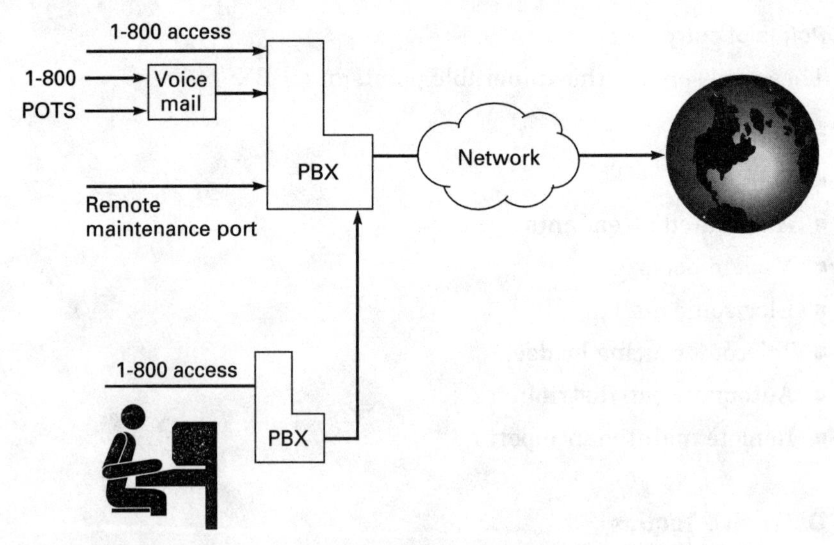

Figure 11.2 Gaining access. (*Source: AT&T.*)

credit cards or be reimbursed from their personal telephone bill. Disallow DISA calls to areas where you do not do business.

2. *Separate internal and external call forwarding privileges* for internal lines. This way, an inbound call cannot be forwarded to an outside line unless authorized. Some internal lines should probably not be able to call forward either internally or externally.

3. *Implement call vectoring.* On selected calls and/or ports, answer with a recorded message, tone(s), or nothing at all. The purpose of not responding is to frustrate attack dialers.

4. *Give operators more information about the calls they transfer.* When they transfer a call, operators should know:

- The trunk the call came in on
- The extension transferring the call
- The line over which the call will go out

Sometimes a hacker will call a company employee and ask to be transferred to a nonexistent employee. The usual response is to transfer the caller to the operator. The hacker then asks for an outside line. The operator should be suspicious if a caller from the *outside* asks for another *outside* line. This should never occur.

Verify that calls being transferred are sent to a legitimate extension in the PBX's database, not to an outside line. In some systems, an inbound caller can get an internal dial tone. By dialing 9, they can gain access to an outside line.

Use multilevel passwords and access authentication. Newer PBXs support at least three password levels. Some also use speech recognition for access to specific functions. If you elect to use speech recognition, do not set the parameters too tightly. A bad connection or a cold could keep a legitimate user out.

Hospitals are particular targets. A caller uses social engineering: he or she poses as a telephone technician and calls the operator, saying that when a third-party call verification request comes in, the operator should say "yes," and that the hospital will not be charged. Of course it is, sometimes for thousands of dollars.

A social engineer calls and asks for a doctor (a real name may be used) and is then transferred several times before again reaching the operator and asking for an outside line. A slew of calls are then made.

Simple rules can prevent much mischief:

- Do not accept collect calls from people claiming to work for the telephone company.

- A telephone company employee would not ask to charge calls to another number.
- Telephone companies do not ask their customers to help with technical problems or entrap criminals.
- Do not accept third-party charges or give an outside line to someone unknown to the operator.

Clues that attacks have occurred

These are good rules not just for hospitals, but for everyone.

All but the smartest hackers leave telltale clues to their presence (see Figs. 11.3 and 11.4):

- *Upon trying to enter the system,* attack dialers leave many incorrect user ID and password pairs.
- *Upon exiting the system,* nonexistent "phantom" extensions will appear as call sources; rarely used access codes and equal access codes will show up in audit trails or call-accounting reports.
- *Abnormally heavy usage* occurs that is not explained otherwise. Sometimes hackers' own greed is the only thing that gives them away. Oddly, after a hacker has succeeded, time changes its allegiance to us, not the hacker. The hacker is now on the defensive, trying to hide from us.
- *Unexplained systems changes* may imply that someone has changed the system illicitly. Attempts to access system programming are indicated by many failed password attempts on the

AT&T Communications — TFEDS — Page 8
Program: I7HG2150 — Fairhaven — 03/01/93 17:34:26
03/01/93 Detail — 800/Orig Security Detail Report — Reten Code: 01R-00300

800 Number NPA NXX-Line	From Number NPA NXX-Line	Connect Start	Date/Time Last	Message	Minutes	Avg Hold
800 525-7133	215 688-6195	0301 1504	0301 1504	1	0	0.00
800 525-7153	215 688-6195	0301 1505	0301 1505	1	0	0.00
800 525-7173	215 688-6195	0301 1505	0301 1505	1	1	1.00
800 525-7193	215 688-6195	0301 1507	0301 1507	1	0	0.00
800 525-7213	215 688-6195	0301 1508	0301 1508	1	0	0.00
800 525-7223	215 688-6195	0301 1509	0301 1509	1	0	0.00
800 525-7233	215 688-6195	0301 1509	0301 1509	1	0	0.00
800 525-7243	215 688-6195	0301 1509	0301 1509	1	0	0.00
800 525-7263	215 688-6195	0301 1510	0301 1510	1	0	0.00

Figure 11.3 A hacker searches for an 800 number. (*Source: AT&T.*)

AT&T Communications		TFEDS				Page 1
Program: I7HG2920		Fairhaven				03/02/93 11:10:56
		High-Fraud Exchanges				Reten Code: 01R-00300
800 Number	**Originating**	**Connect Date/Time**	**Range**	**Calls**	**Minutes**	**Avg Hold**
800 237-[illegible]	212 781-6517	03/02 09:21	03/02 09:42			
800# Total		03/02 09:21	03/02 09:42			
800 241-[illegible]	212 690-0025	03/02 09:14	03/02 10:11			
800# Total		03/02 09:14	03/02 10:11			
800 345-[illegible]	212 281-5786	03/02 07:52	03/02 10:18	10	262	26.20
800# Total		03/02 07:52	03/02 10:18	10	262	26.20
800 544-[illegible]	212 568-8058	03/02 09:48	03/02 10:17			
800# Total		03/02 09:48	03/02 10:17			
800 633-[illegible]	212-304-3589	03/02 09:13	03/02 10:14			
800# Total		03/02 09:13	03/02 10:14			

Figure 11.4 A hacker finds an 800 number. (*Source: AT&T.*)

PBX maintenance port or a computer attached to it. The solution is to monitor external access to the port and track the problem to its source using telephone company resources, automatic number identification, or caller ID.

Going dark

One option is to turn off all remote outside access altogether and disconnect any maintenance ports. Hackers call this "going dark." Going dark amounts to blackmail before the fact. It damages your organization by denying employees the benefit and competitive advantages of DISA and other services such as conference calling. A lost deal could cost much more than a hacking incident that might never occur.

Instead of going dark, implement access controls and auditing. Use the recommendations above. Talk to your PBX vendor(s) and your carrier(s). They all have a vested interest in preventing toll fraud. It is bad for customer relations and they do not want it any more than you do. Each carrier uses their toll-fraud detection and prevention plan as a competitive value-added feature of their network. Make use of them.

PBX Vendor Responses

The PBX manufacturers have been placed on the defensive. Their responses include the following:

- Users must take advantage of the tools built into PBXs before they can legitimately fault the vendor.
- Use warning labels, security manuals, and training seminars to make the problem and its solutions visible.
- Users should be required to choose their own passwords, rather than using determined ones.
- Multiple levels of passwords for DISA ports should be set up. Vendor InteCom lets users change their passwords as often as every five minutes, and is working on voice recognition as an authentication method.
- Users should implement call-back from PBX maintenance ports to a predetermined number, not necessarily the number that called (see Fig. 11.5).
- Some vendors, notably Northern Telecom, Inc., have incorporated antihacker software or hardware. Northern Telecom's product, called Meridian Safe, is based on purchased Xiox Corp. technology.

Some users have been critical of vendors who have strengthened weak switches against attack and then charged the customer for that improvement. For example, some telephones show the caller the number they have dialed. They will also show a user's password into voice mail. New software is needed to blank the display during password entry.

Users have also claimed that maintenance organizations should bear some security responsibility as well and be specifically recognized in the FCC's rulings. The logic here is that maintenance port access codes can be stolen from these firms.

Voice-processing vendors such as Octel have added multilevel passwords and a self-destruct feature that lets users delete all messages in their mailbox if they forget their password. The system administrator must delete and then add the mailbox again. This prevents a hacker from acting as an administrator and looking in a user's mailbox. New mailboxes are assigned a random 6-digit number by the system. When a mailbox is accessed, it plays back the time of the last access as an audit check for the user.

Fighting Back

Clearly, outside access ports should have additional barriers placed before them. There are many such products on the market, such as LeeMah Data Security's TraqNet 2000-V.

DISA Port Protection

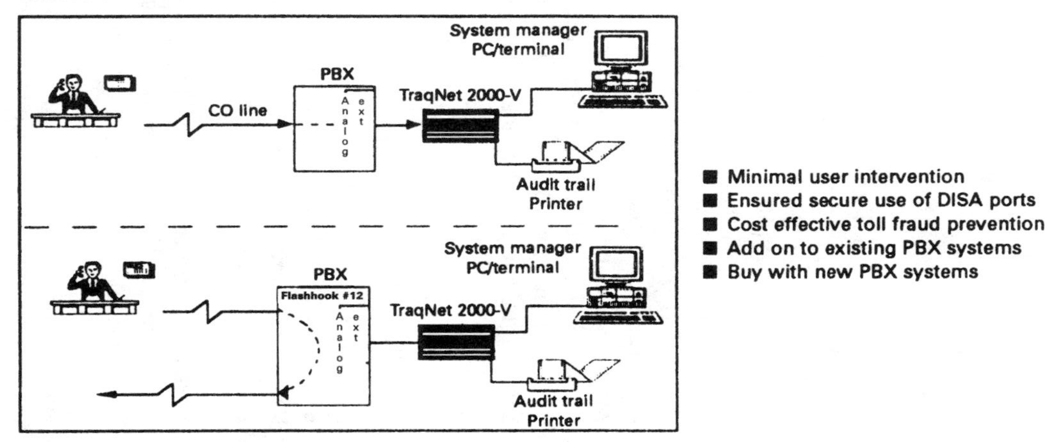

- Minimal user intervention
- Ensured secure use of DISA ports
- Cost effective toll fraud prevention
- Add on to existing PBX systems
- Buy with new PBX systems

Maintenance Port Protection

PBX
Maint. port
RS232
Modem
Remote user
InfoKey or InfoDisc
Modem
RJ11
TraqNet 2000-V

- Transparent secure access
- Detailed audit trail of all use

Voice Mail Protection

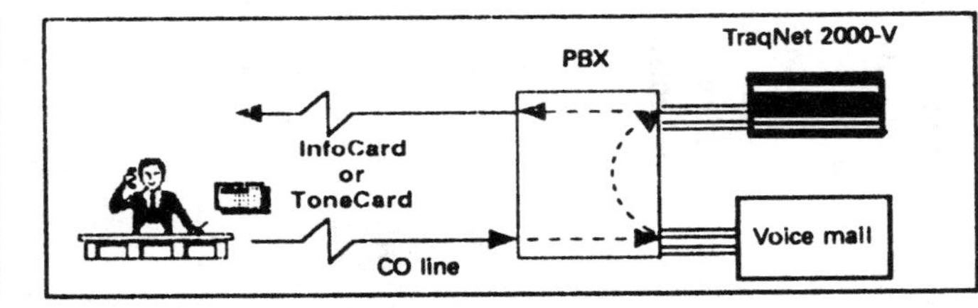

- Prevents unauthorized trunk access
- Minimal user intervention

Figure 11.5 Port protection methods. (*Source: LeeMah Data Security Corp.*)

Call-accounting software should be in place and secure. Software that looks for hacking patterns can be a great timesaver. Be alert for hackers' tricks and bring them to the attention of management immediately if you suspect trouble.

Management's moral support is often strong, but such moralizing often does not translate into dollar support. The following examples illustrate how some of the world's most disciplined organizations have been victimized.

Pitney Bowes in Danbury, Conn. was struck for thousands of dollars in toll fraud, starting with the night security guard who used voice mail to get to a toll-free number and call his relatives in South America. Outside crackers broke into Pitney Bowes's data network as well.

The *United Nations* was hit. *NASA* lost $8 million in telecommunications fraud. Unbelievably, the *Drug Enforcement Agency* was cracked by drug dealers and *AT&T Bell Laboratories'* own networks were penetrated on the average every other day via their Internet link.

Dodger Industries lost $35,000 via their 800 number. Other victims include the *New York State Welfare Department, Pacific Mutual Insurance,* and *Avon Cosmetics* (losses undisclosed).

Mitsubishi International Corp. settled a multimillion-dollar dispute with AT&T out of court. Terms of the settlement were never disclosed. Similarly, *Shiva Corp.* was struck by a hacker but had a toll-fraud monitoring service that caught the fraud quickly, as we will see.

Jiffy Lube

In a landmark case, the U.S. District Court of Maryland decided *AT&T v. Jiffy Lube International* against Jiffy Lube. The matter involved some $55,700 plus 8 percent compound interest since 1988 for unauthorized calls placed from Jiffy Lube's PBX into the international network.

The court decision stated in part that "the tariff squarely places responsibility upon a customer, such as Jiffy Lube, for calls, whether or not authorized, that 'originated' at the customer's number...at Jiffy Lube's PBX."

The use of DISA access influenced the court. In doing so, Jiffy Lube "created the vehicle and mechanism by which those [abusive] long-distance calls became possible." From the legal view, said the judge, illegal calls are the same as those made by a burglar inside a users' building, which the user would have to pay for as well.

The case was complicated by the fact that the inbound 800 access was provided by MCI and the outbound calls were carried by AT&T. The judge felt that AT&T should not be expected to protect a customer "which lawfully opts to contract with a competitor."

Analysis. This case took place before all of the Big Three carriers (AT&T, Sprint, and MCI) instituted their toll-fraud detection systems. However, it does illustrate that users victimized by toll fraud are almost always losers in court. It also illustrates the fact that toll-fraud laws on the books need to be strengthened.

Hospitals

Hospitals are increasingly targeted as toll-fraud victims. Here is a description of two scams.

Calling the hospital's main number, a hacker poses as a telephone company technician or a security officer. The hacker asks the receptionist to help in an investigation by saying "yes" when an "operator" calls to verify third-party charges. The hacker assures the operator that the hospital will not be charged. But of course, it will. These calls are usually international and run into thousands of dollars.

In the second scam, "Dr. Jones" makes a collect call to the hospital and asks to be transferred to "Dr. Smith." (These are often real names as they are easy to get.) "Dr. Jones" may be transferred several more times before winding up back at the receptionist. Then he asks for an outside line. Once given, unlimited calls can result.

These real scenarios can be prevented without resorting to high-technology barriers:

- Never accept a collect call from anyone calling themselves a "phone company employee." Employees never have reason to ask such a thing, or to charge calls to a third number.
- Telephone company employees never ask customers to help trap a crook or assist with any maintenance procedure.
- Never give an outside line to and never accept third-party charges from an unknown person.

Federal Communications Commission Rulings and Congressional Action

To date, the FCC has taken no significant action in the toll-fraud issue. They did hold a one-day hearing to help determine what, if any, role they should take in these matters. Toll-fraud experts, vendors, and users testified. The FCC took no action, and so left the matter to the courts and Congress.

While the FCC has no legal responsibility to prosecute toll fraud, it is proposing new rules to fight toll fraud:

- For pay telephones, the FCC is proposing that providers who take reasonable steps to limit their vulnerability to toll fraud should not be required to pay bills resulting from fraud.
- For cellular systems, the FCC proposed:

  Technical specifications to prevent tampering with the electronic serial number in mobile equipment.

  To find out if there is sufficient incentive for industry to provide fraud prevention in customer premise equipment.
- For calling cards the FCC would require:

  Long-distance carriers to provide data to LECs that allow them to develop calling pattern profiles.

  A rule for allocating liability for toll losses.
- For equipment vendors:

  Vendors would have to provide warnings and data on how to limit fraud risk. (This is already being done without FCC prodding.)

  Vendors should explain the risk of failing to change default passwords and codes. (This is also being done.)

On balance, the FCC wants to eliminate fraud using the resources of all parties involved. It wants to define what the responsibilities are of each of the parties and then hold them responsible for them. Failure to meet that responsibility would cause a party to be responsible for some or all of the financial loss accruing from a fraud incident. However, the FCC is generally inclined to lay more responsibility on the carriers and vendors rather than the customers. Furthermore, carrier tariffs limit their liability for toll fraud. The FCC intends to reexamine those tariffs and the underlying policy for having them.

Some members of the user community want a cap on toll-fraud losses in the same way the Truth in Lending Act limits the liability for calling cards to $50. For toll fraud, one user suggested the first four hours or $500 of illicit calls. If the carrier failed to notify the user, then it would absorb 50 percent of the remaining loss. Another user suggested splitting the loss equally between user, LEC, interexchange carrier (IEC), and vendor.

In 1992, Representative Edward Markey (D–Mass.) introduced The Telephone Toll Fraud Remedies Act of 1992. The bill would have required carriers to add security to the public network to detect and prevent fraud and would have given users a way to block international calls on PBXs.

Vendors of PBXs, automated attendants, and automatic call distributors would have been required to issue warnings to owners notifying them of the possibility of toll abuse, and requiring vendors to give advice on how to thwart it. Many PBX vendors are doing this now, even without a legal mandate. In case of abuse, arbitration would have been required instead of costly legal battles. Perhaps most significantly, users would have been held liable for only a third of any disputed amount. This bill died when Congress adjourned for the year.

Call-Accounting and Authentication Products

You can preemptively install call-accounting software and/or products requiring additional authentication. Many companies offer such products including the following.

Xiox Corporation of Burlingame, Calif. offers three products to detect toll abuse, as shown in Fig. 11.6. Uniquely, the Hacker Preventer uses artificial intelligence to sense deviations from a user's profile and asks for progressively more authentication. If further authentication is lacking, it terminates the call. To reduce the likelihood of hacker access, Hacker Preventer runs on a separate PC hardware platform and costs between $10,000 and $28,000.

The company also sells Hacker Tracker, call-accounting software that uses the same station message detail recording data that is

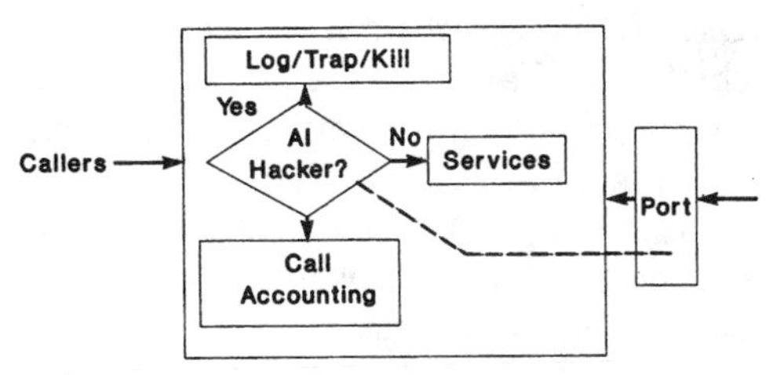

Figure 11.6 Locking the PBX (Xiox method).

used in a call-accounting system. Hacker Tracker looks for international calls, especially to area code 809, monitors voice mail activity, and contains a dialing-recognition template to set traps or alarms for specific types of hacker attacks. Like Hacker Preventer, the software runs on an IBM-compatible PC-AT, PS/2, or 386/486 PC.

Hacker Deadbolt locks up remote maintenance and test ports for PBXs, voice mail systems, and the like. It can be purchased as a standalone device or built into the Hacker Preventer platform. Deadbolt also records attempts to access maintenance ports. Security options such as extra passwords, forced waits between call attempts, alarms, and artificial-intelligence-coupled monitoring add back door security. Hacker Deadbolt requires physical (RJ-11) connection to the maintenance or test port to be protected.

The Xiox series of products is also marketed by MCI and the regional Bell holding companies and is sometimes purchased by other hardware vendors and sold as their own equipment as in Northern Telecom's Meridian Switch. Xiox products are generally considered to be among the more sophisticated and powerful products available.

For example, the Travelers Corp. insurance company offers a 25 percent discount on their toll-fraud insurance policies when Xiox equipment is used. In addition, Xiox offers insurance underwritten by Great American Insurance Co. to insulate the user from loss (to a specified limit and with a deductible).

Account-A-Call

Account-A-Call uses 25 or so internally developed exception algorithms to detect toll fraud. Patterns suggesting fraud or abuse, such as excessive or overlapping domestic code and credit card use, are noted. International calls are a particular red flag. Their reports contain both graphic and standard tabular formats for management review.

Account-A-Call's standard method of operation is shown in Fig. 11.7. The Tadpoll data collector is connected to the PBX's station message detail recording (SMDR) port, and gathers call detail information from the PBX. The Tadpoll also does its own real-time checking and sends an alert to the client and Account-A-Call's central monitoring office if it sees unusual activity.

Normally, after hours, Account-A-Call (AAC) polls the device and uploads the call data to the central monitoring office. Host machines then reduce the data into a form presentable to an analyst. The analyst notes unusual patterns if any. If fraud or abuse is suspected, special reports are prepared and client management

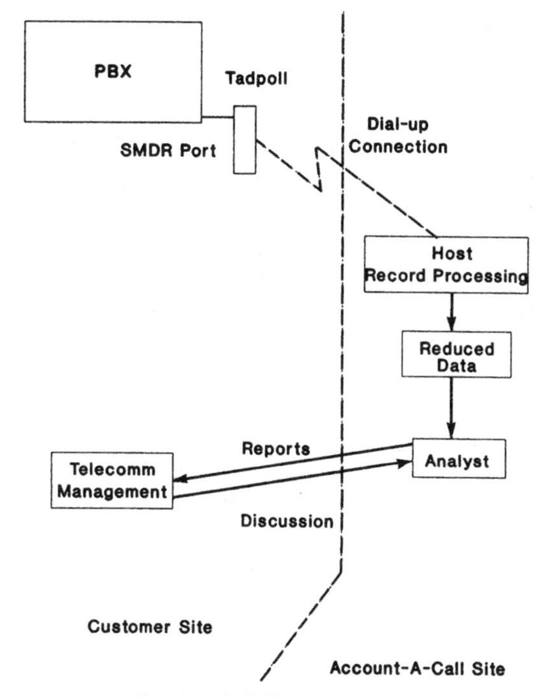

Figure 11.7. Account-A-Call's approach.

informed. AAC also prepares standard monthly reports of call activity. The system worked in the case of Shiva Corp., a high-technology manufacturer of Apple Macintosh communications products located in Cambridge, Mass.

The Shiva incident. In 1992, a caller claiming to be a New England Telephone technician called Shiva's telecommunications manager. He said that he would be performing trunk testing on the company's lines. The "technician" said that the calls would appear on the firm's telephone bill as collect calls, but Shiva would not be charged for them. This fairy tale alone should have been enough to trigger suspicions. No reputable telephone company employee would *ever* make such a request.

Although his suspicions should have been raised, the telecommunications manager instructed the person who runs their Northern Telecom Meridian I PBX to expect the calls. The "technician" then began to make calls. What the hacker did not know was that Shiva had previously engaged Account-A-Call to monitor their PBX activity. A few days after the calls began, an Account-A-Call analyst noted a pattern of direct trunk access 800 calls. These are calls in which an inbound caller uses an inbound 800 number for access to an outbound switch trunk. The analyst also noted trunk numbers in call record extension fields. These numbers indicated a caller making many outbound calls on a particular switch trunk. Long-duration calls, another clue, were also recorded.

Account-A-Call called Shiva, who then called New England Telephone (now NYNEX). New England Telephone traced the calls to a convict already behind bars in a Boston-area prison. (Prison telephone numbers are considered high-fraud sources and appear in many antifraud databases.)

In all, there were 100 to 150 calls involved, not many as toll fraud goes, but with usage as high as 20 hours a day. Had it not been detected early, it could have been a major loss. As Shiva's telecommunications manager said, "If we hadn't had [Account-A-Call's service], this could have developed into a cost of critical proportions. *It certainly underscores the value of detecting toll fraud early*" (italics added). We might add that the experience also underscores the need to be aware of attempts at social engineering as a means of toll fraud.

Though not the usual practice for a carrier, New England Telephone credited Shiva for the calls before they were billed...ironically, just as the hacker claimed.

Blue Cross and Blue Shield. In a novel arrangement, Blue Cross and Blue Shield in Chicago set up a system whereby call detail records from their carriers, AT&T and MCI, are sent directly to Account-A-Call. If suspicious activity is noted, AAC calls Blue Cross.

Currently, the information is passed on tape to Account-A-Call, who prepares weekly and monthly reports for Blue Cross and Blue Shield management. AT&T provides exception reports for its Software Defined Network and MCI does the same for its Virtual Private Network and 800 services. This arrangement is cheaper for Blue Cross and Blue Shield than buying and administering their own call-accounting software.

Speaker Verification Systems

Wye Technologies of Annapolis, Md. offers a speaker verification system called VVS. Beside authenticating access, VVS is a platform for a variety of concurrent voice applications, including order entry, call center systems, messaging, and media conversion. VVS works with all major telephone switches.

Moscom Corp. of Pittsford, N.Y. offers a product that performs voiceprint authentication called VoiceLock. ("I'm sorry, but your voice is unacceptable"). Their associated Hacker Tracker (note that this name duplicates that of a Xiox product) software notes unauthorized calls and abnormal calling patterns in the vendor's Discovery/1 PC-based call accounting system.

Telco Research Corp. of Nashville, Tenn. connects its product, Tru Real-Time Toll Fraud Detector to a PBX's SMDR port. It detects suspicious activity as defined by the user and generates alarms. Excessive authorized code use, too many international calls, too many calls by trunk group or circuit, calls after business hours, short-duration calls, certain dialed numbers, power loss, and data loss are all flagged. The product includes an EIA-232 serial port and an internal 2,400 bit/s modem.

Microframe of Edison, N.J. offers a PBX antifraud product that works in real time and permits user-defined alarms. It also offers data network security products.

Telecom Services Limited, Inc. is a New York City–based service bureau. They help to negotiate loss restitution with carriers in case you experience a hacker-induced loss.

Preventing Cellular Fraud

Cellular systems are by no means immune from fraud. Of the $4 billion or so lost annually, about $365 to $600 million of that is due to cellular fraud. Every day in the cellular industry over $1 million is lost to fraud. On that same day just *one* person is apprehended for fraud. This is going on even as 17,000 new subscribers join the network daily. They now get to pay their share of some unknown hacker's calls all over the world. Fair?

In response, *GTE, Coral Systems,* and a *Pacific Telesis and Electronic Data Systems* team have each developed software used by cellular carriers to prevent cellular crime.

Coral Systems' product is called *FraudBuster,* an artificial-intelligence, workstation-based product that watches the calling behavior of users over time. It takes between 14 and 40 days to develop an adequate user pattern. By use of artificial intelligence, it is possible to filter out the few likely fraud cases from the many that are examined. FraudBuster worked well in a June, 1993 test. It has since been deployed in Puerto Rico and in several Bell company cellular networks. It is now being deployed by AirTouch Communications as well.

The system can also detect cell phone cloning, where two phones have the same number but are calling from different places. These numbers can be recorded easily by anyone with a high-frequency scanner or even with a modified UHF television.

By constantly reprogramming or "tumbling" the electronic serial number (ESN) in a cellular phone with another ESN, it is possible to stay a step ahead of the cellular carrier. FraudBuster detects the ruse by checking the Home Locator Register, a database of legitimate users. If not there, the call is cut off. However, when one carrier detects the problem and invalidates the telephone number, the crook often simply mails the phone to another state and another cellular system operator where it will start a new but hopefully brief life.

Cellular criminals also buy cloned phones programmed with stolen serial and telephone number combinations. Unlike cellular tumbling, clone fraud cannot be detected by comparing serial and telephone number combinations against the service provider's database, since the telephone numbers are legitimate. This means that their calls are billed to *us*.

To stop this, PacTel Cellular in Los Angeles is testing a product called PhonePrint, built by ESL/TRW. It compares the unique signal frequency patterns of the cellular phone making the call to a pattern on record for that device. Accordingly, PhonePrint can detect a mismatch between a device and a telephone number. It will stop cloning by blocking calls where the pattern and number do not agree. AirTouch and Coral are tying FraudBuster to TRW's PhonePrint to absolutely validate a user as legitimate or not. It is estimated that 85 to 90 percent of this problem can be solved this way.

Carriers have been reluctant to install such software because the cellular market developed slowly. Already deep in debt, the carriers elected to (make us) live with the problem in the form of higher rates. Faced with new investments for digital upgrades and Cellular Digital Packet Data, security took a back seat as it often does when push comes to shove.

PhonePrint and a competitive product, PacTel and EDS's *Clone Guard,* is installed in PacTel's cellular network. A functionally similar product, GTE *Clone Detector,* is installed in SNET Cellular's New Haven, Conn. area network. Does *your* cellular provider have such protection for *you?*

Yes or no, there are some things you can do:

- Find out what if any security your carrier employs. For new bids, make it a mandatory requirement.

- Recognize that security is weak.
- Make the people who use the phones check their bills every month.
- Understand that every call is public. Do not disclose credit card numbers or any information you deem sensitive on a cellular phone. Remember what happened to Prince Charles (among others).
- If justified, use a voice encryption device. This device mounts between the handset and the phone base using the modular clips we know so well.

User Information Exchange

If hackers exchange information then so should the user community. Users do so through the various user groups whose nuclei are the switch vendors: e.g., the Unauthorized Access Committee of the International SL-1 Users Association, a group of Northern Telecom, Inc. PBX users. You can also contact the Telecommunications Council of Multinationals, the Security Committee of the AT&T National Definity Users Group, the International Communications Association, the Telecommunications Association, and regional organizations such as the Communications Managers Association (in the northeast) and the Southeast Telecommunications Association.

Contact your switch vendors' users group to see if they have a toll-fraud special interest group. Almost all do. Your switch vendor can put you in touch with the user's group.

The North American Telecommunications Association (NATA)

Contact the North American Telecommunications Association. NATA offers a 52-page report titled "*Remote Access Toll Fraud: Detection and Protection*" to help users repel PBX violators.[2]

The Toll-Fraud Prevention Committee (TFPC)

The TFPC is also based in Washington, D.C. and consists of the Big Three carriers, the regional Bell operating companies, independent telephone companies, and many smaller companies across the country. TFPC is mainly a technical group.

Its prime focus has been to stamp out international fraud and illegal bill to third-party calls (credit cards, mainly). This group finds and voluntarily implements technical solutions on an individual basis. Its recommendations are not subject to regulatory approval.

One ruse TFPC targeted was fraudulent calls made from overseas to a pay phone in the United States. Some committee members elected to use a single specific digit that appears on the screen in front of an operator just before the last four digits called. This means that the number is a pay phone. This prompts the operator to inquire further before completing the call.

Another problem is "shoulder surfing," where someone looks over your shoulder and notes your credit card number. Bare eyes, binoculars, and even video cameras with telephoto lenses are used. Occasionally the phone itself is tapped and the tones are simply recorded. Once the numbers are stolen, a call is made to a remote pay phone. Three-way calling then bridges the call to its destination.

A recent television program showed New York City police and AT&T Corporate Security working together to arrest shoulder surfers. When an illegal call was made, it was detected by AT&T and the perpetrators were arrested on the spot.

This strategy was so successful that AT&T has formed an official investigation team to work with customers and law enforcement when the "stakeout" software in Definity switches detects a hacker trying to obtain dial tone. The service will become a part of its PBX customers' standard maintenance contracts. It is important to catch the original hacker, because subsequent calls may be made by others, and may be placed from anywhere.

A disadvantage of this strategy is that if effective, small dollar losses are involved. This lessens the motivation to prosecute or prosecute aggressively. Furthermore, crawling around inside a PBX is not a criminal act. If a hacker finds a dial tone and then passes it to someone else to make calls with, the hacker himself or herself is unlikely to be prosecuted.

Often, illegal calls are routed through small towns to help keep them from being located. Logically, AT&T and law enforcement concentrate on the big cities with large immigrant populations, such as Los Angeles, Miami, and New York.

What you can do

Indirectly, the TFPC has started a marketing campaign to teach people how to keep their numbers from being stolen. The rules are as follows:

- Do not speak the number you seek loudly.
- Keep an eye out for people with binoculars, video cameras, or inquisitive stares.
- Use your other hand or a piece of paper to shield the number pad while you enter your card number. Some telephones have built-in shields. These do not always work and so can foster a false sense of security. Crooks can also attach wires to pay phones to record tones and then play them back to bill to your card.
- Review each telephone bill received. The telephone company will tell you to whom you made a call if you give them the number. Carriers also monitor credit card usage for fraud automatically as part of the credit card service.

Not all TFPC members install all of the committee's recommendations, especially the smaller ones, so implementation may be spotty. The committee meets four times a year and is open only to members due to the sensitive nature of the discussions.

Carrier Toll-Fraud Monitoring Programs

The three largest interexchange carriers all now have toll-fraud monitoring and prevention programs. AT&T's is NetPROTECT^{SM}; Sprint's is SprintGuard, and MCI's is Detect. They are all basically similar in structure and purpose. We discuss AT&T's and Sprint's programs as case examples.

AT&T's NetPROTECT

NetPROTECT services are designed to prevent, detect, and control toll fraud originating from other than the customer's premises. NetPROTECT services are offered for customer premise equipment on the U.S. mainland, Hawaii, Puerto Rico, and the U.S. Virgin Islands. A business service only, subscribers must use one of 16 AT&T products, broadly, 800 services, Megacom, WATS, Software Defined Network, or business international long distance.

Calling card and operator-assisted calls are not covered by NetPROTECT, although AT&T does monitor card calls for possible fraud and notifies the cardholder.

NetPROTECT is offered at three levels:

- *Basic Service* provides a $25,000 liability limit which is halved if the client detects and reports it first. This applies to one incident of fraud in a 12-month consecutive period.

- Customers eligible for Basic Service have the option to upgrade to *Plus Service*. Under Plus Service, liability is limited to $7,500 and is again halved if the customer reports it first.
- *Premium Service* downgrades customer liability to zero for an additional charge based on locations covered.

Virtually all business subscribers who have on-site PBX or key systems receive NetPROTECT Basic Service at no additional charge. The protection comes from the toll-fraud early detection system, or TFEDS. The service is available 7 days a week, 24 hours a day, and operates in near real time. Security-trained personnel are always available at (800) 821-8235. Credit card, domestic 800, and 1 + international calls to over 160 countries, especially those with high-fraud history, are monitored.

How NetPROTECT works. Inbound calls from a unique automatic number identification (ANI) might indicate a hacker. These calls can be blocked by AT&T. Fraudulent outbound calls can be blocked as well with the subscriber's permission. In fact, the carrier (by FCC ruling) has thrown hackers off 800 service as soon as abuse is detected, even without subscriber permission. AT&T has done so *over 1,000 times* since August, 1992. Unusual international calls are also noted. AT&T tries to notify the subscriber so they can secure their system.

AT&T Corporate Security will work with a subscriber to diagnose toll fraud, find out if it originates in the customer premise equipment, and then block access. AT&T will not only help capture the criminal, but they will help gather data needed to prosecute the offending hacker. Their corporate security also assists in stopping annoying 800 calls through its 800 Service Annoyance Call Bureau. In fact, AT&T now impounds misused 800 numbers to prevent abusers from importing them to another carriers' network.

Education is part of the program through fraud control seminars and employee awareness training program development. AT&T also educates local law enforcement officials and lobbies states for stronger antihacking statutes.

The program has met with some success. AT&T trained 14,000 customers last year through its Fraud Awareness and Prevention Seminars, and average toll fraud is down 75 percent in the first two years of NetPROTECT's operation. Some 2,500 hackers have been thrown off the network and 1,000 customers a month are called regarding suspicious calling patterns.

Despite all the above, all calls made are the responsibility of the subscriber, no matter what.

NetPROTECT Plus. Plus Service includes all of the Basic Service features and is intended for small- and medium-sized businesses with call volumes under $20,000 per month. It covers AT&T Business Long Distance and 800 services.

Customers pay a one-time installation charge of $100 per location, with a monthly charge of $10 per location thereafter.

All services and phone numbers at a single site are covered. Other common carrier ANIs are monitored as well if 10288 is dialed to place outgoing calls. Even a simple key system is covered.

Up to three customer contacts can be named to be called (on a 7-day, 24-hours-a-day basis) in case of a suspicious incident. With permission, AT&T will block all outbound calls until the premises system is secured.

Plus Service is a tariffed service and is available in areas where AT&T provides customer call detail recording, about 92 percent of all U.S. business sites. These records are analyzed and international calling patterns kept on file.

Premium Service. Premium Service includes all the Plus features yet goes further to build in a zero initial liability limit. From the first fraudulent call until AT&T notifies the subscriber of a problem, the customer is fully indemnified. For the next two hours, the customer has no liability while the switch is secured. After that, the customer's liability resumes. Due to its sophistication, Premium Service detects toll fraud some five to six times faster than customers who have only Basic service.

For this service, there are some fairly stiff requirements. The customer must:

- Have any vendors' PBX or key system
- Have AT&T Outbound Services
- Have no 800 inbound service or 30 percent (limited) AT&T inbound service per customer premise equipment (CPE) (e.g., PBX, ACD)
- Limit access to remote ports
- Block call transfer from remote ports
- Protect inward system access (DISA)
- Have call accounting[3]
- Provide a list of originating telephone numbers
- Provide a contact list of three people
- Have a crisis intervention plan on file with AT&T

- Provide a profile of calling patterns

The customer has zero initial liability instead of a $25,000 cap. There are additional customer requirements and a location-based installation fee structure: $300 for the first 100 locations; $250 for the next 200 locations, and $200 for all remaining locations. Monthly service is also three tiered: $200 per month for the first 100 locations; $150 monthly for the next 100 locations, and $100 a month for any remaining locations.

Added requirements include all-AT&T 800 inbound traffic per CPE, a longer DISA password, and an AT&T Security review of the customer's CPE.

Teleconferencing service. NetProtect protects users against hacking via special teleconferencing service features.

Other services. AT&T offers a variety of other security services for SDN in particular, including use of screening and caller groups, location screening, and authorization codes. For users of AT&T 800 Services Advanced Features, routing-on-demand changes now require a user access code and a personal identification number. AT&T also offers Hacker Tracker, a security audit service, consulting, and a security handbook. For more information, call AT&T at 1-800-NET-SAFE.

ISDN Caller Information

Using ISDN's ANI, mass marketers can identify whether a call to order merchandise is coming from a pay phone, a favorite hacker trick, or a cellular phone. In the words of an AT&T district manager, mass marketers "...can use [ANI] to make sure they're not sending $15,000 worth of products to an inmate that called in from Attica [a high-security prison in New York State]."

In a marketer's automatic call distributor, the ANI would set a flag warning the marketer that the call is from a pay phone or from a prison. The system can also be used to generate demographics to help marketers tune their pitches. For instance, orders via cellular phone imply that these callers have greater incomes and mobility.

AT&T is currently the only carrier to offer what they term *Information Indicator,* but MCI and Sprint both offer ISDN and could follow suit. The Information Indicator (II) digits can identify locations that have high fraud potential. II digits are sent in the D channel of the 23B + D Primary Rate Interface using the Q.931 protocol. A typical charge is a penny per ANI or II digits.

Discussion. The Premium Service is very expensive in terms of qualifying requirements. At some point, one cannot justify the combined expense of paying a carrier plus having an extensive amount of equipment and software. With any carrier plan, it is important to have weekend service. That is the hackers' busiest time.

SprintGUARD

SprintGUARD CPE Security Support Service is a no-charge service designed to:

- Minimize user exposure. Sprint claims only a $950 per-incident loss compared to a national average of $45,000.
- Maximize network availability by deterring and detecting network abusers.
- Minimize the Sprint network's exposure. Sprint claims that CPE fraud on its network was under 1 percent of the 1992 industry estimate for total CPE fraud losses. (Recall, however, the remarks at the beginning of this chapter concerning the range of loss estimates.)

The service consists of:

- Selective monitoring and traffic analysis of inbound domestic 800 and international toll-free service. Outbound international and area code 809 calls are also watched. Customers are notified of suspicious activity and corrective recommendations are made.
- Hacker bulletin board monitoring.
- 24-hour security support at Sprint's Corporate Security Department: (800) 877-7330, including investigation.
- Security education.
- Consulting services.
- Emergency call tracing (same 800 number as above).
- Liaison with other domestic and international carriers.
- Law enforcement and prosecution support.

Sprint also offers a higher-level service called SprintGUARD Plus^{SM}. This program provides financial protection against fraud. Under this plan, Sprint takes responsibility for any CPE fraud incident exceeding $10,000 in fraud at a CPE site. Call types, such as

800 and international calls, are the same as with SprintGUARD. However, monitoring and analysis of these calls are more comprehensive and take priority over basic SprintGUARD activity.

Should suspected fraud occur, Sprint will send a call detail report via facsimile to the customer. Call history going back to the customer's last invoice can be provided within five business days of detection, to head off unexpected bills on receipt of the next invoice. The entire matter is handled by a security support manager assigned to assist and recommend corrections during a fraud incident.

Periodically, SprintGUARD Plus customers receive bulletins to stay up to date on current events in fraud detection. These customers also have the benefit of the CPE Security Support Service.

Local Exchange Carriers

Traditionally, local exchange carriers (LECs) have not taken a strong position on interexchange carrier (IEC) fraud because they will be paid by the IECs regardless of whether the IECs are paid or not. One exception is U.S. West, Inc.'s Customized Fraud Management Service, intended to track attempts to place operator-assisted and calling-card calls. The system is intended to help the interexchange carriers and others detect fraud and decrease the chance of a customer becoming a victim.

Two kinds of monitoring are included:

- Velocity checking tracks call attempts over time.
- Screening monitors calls to specific telephone numbers.

Data will be derived by tracking queries to U.S. West's database, rather than monitoring calls directly. Reports based on these statistics will be given to customers periodically.

For example, if a caller wishes to bill a call to a third number, the IEC can verify in the U.S. West database that this is allowed. The IEC would pay about a half-cent per call for this service, while the customer would also pay for the service: perhaps to both the LEC and the IEC. The benefit is that the LEC can monitor traffic going or coming from any IEC, while IECs can examine only their own traffic. Consequently, an IEC can see call attempts from all IECs used by the customer and use that information to decide on whether to bill the customer for calls believed to be fraudulent. Perhaps most importantly, a real-time check may prevent the fraud from taking place, and so circumvent the entire problem.

TABLE 11.1 A Toll-Fraud Prevention Program

- ✓ Constant vigilance over toll, credit card and employee usage. This includes being receptive to things employees notice.
- ✓ Employee awareness and training is your *best* tool.
- ✓ Secure inbound DISA and voice mail ports.
- ✓ Set call-forwarding privileges carefully and selectively.
- ✓ Use call-accounting software.
- ✓ Lock up all maintenance ports with additional barrier products.
- ✓ Exchange information with professional peers: NATA; PBX special interest groups.
- ✓ Teach users to guard their credit card numbers.
- ✓ Use your carrier's toll-fraud monitoring service.
- ✓ Never, never let your guard down.

Employee Abuse

Some employees regard the use of the telephone as a perk. Those who do need some serious attitude adjustment. No one will mind an occasional call home to check on a sick child or a call to say they will be home late for dinner. The problem is the abuser who calls all his or her relatives in the far corners of the world from the office during lunchtime (at daytime toll rates, too).

Many call-accounting packages are used to bill departments for their share of telephone costs, so department managers quickly find out who is abusing the phones. There is a potential problem if the abuser uses someone else's phone. Sooner or later, the abuser will be caught.

Some companies use the Tadpoll device connected to the PBX to collect accounting data that are then turned into deductions from employee paychecks. IBM does this. Other companies demur using the Big Brother approach, and indeed, it can be counterproductive if employees perceive the discipline as oppressive. This cultural decision can only be made by management in light of company philosophy. In the next chapter, we extend this discussion: How much security is enough?

Summary Checklist

Table 11.2 summarizes the steps to take to lock up your voice system.

TABLE 11.2 How to Lock Up Your Voice System

Item	Action
DISA ports	☐ Change passwords on delivery ☐ Employ multilayer protection software such as Secure System Access Line (Procter & Associates, Redmond, Wash.)
Voice mail	☐ Six-digit minimum password length ☐ It should never match last six digits of the phone number ☐ Never use repeated (111111) or consecutive (678901) numbers ☐ Lock up administration ports using a port protection device ☐ Lock the caller out after three attempts at password validation Multiple lockouts imply an organized attack; increase your vigilance
800 lines used with automated attendants	☐ Prevent voice mail access, trunk-to-trunk transfers, and other access ☐ Divert other calls to a trained operator ☐ Program the attendant to block unwanted traffic after hours
Calling cards	☐ Warn employees about shoulder surfing ☐ Use voice recognition (e.g., Sprint) ☐ Your toll-fraud protection program should cover such theft
Call forwarding	☐ Avoid using features that use trunk-to-trunk connections via class-of-service restriction tables
700 calling	☐ 700 conference number dialing should be turned off; test periodically that this is so
900 or 976 calls	☐ Prevent dialing to 900 or 976 (calling party pays) and local look-alike exchanges to keep hackers from using them
Carrier dial-through	Hacker identifies the victim's chosen carrier; then tries direct-access codes for other carriers ☐ Check your bills carefully ☐ Install hacker-detection software
Clip-on fraud	Hacker taps into your line, may be miles away ☐ Maintain physical security (lock everything) ☐ Use documentation to prove that fraudulent calls originated from other than your premises
Social engineering	☐ Counteract with awareness training ☐ Use fraud-monitoring software ☐ Place PPDs on LANs
Your general strategy	☐ Monitor your systems constantly ☐ Practice disciplined management ☐ Use the best technical countermeasures you can afford: Even if you install them bit-by-bit, you have a *strategy*

SOURCE: Abstracted from Bernie R. Milligan, *Gateways to Toll Fraud,* Infosecurity News (1994).

Review

- Data hackers steal information; phone hackers steal service.
- Toll fraud is the unauthorized use of toll telephone service, whether by internal but mostly by external sources.
- PBX, credit card fraud, and employee abuse are the basic forms of toll fraud.
- Conservative estimates place toll fraud as a $1 to 5 billion business loss annually.
- Much toll fraud takes place by access to outside lines via voice mail or direct inward system access ports.
- Hacking is indicated by many short, incomplete calls; abuse is indicated by long calls, especially to foreign countries.
- The PBX, ACD, or voice mail system is also vulnerable on maintenance ports.
- Defense tactics mean using all security features in your equipment and removing all weak passwords including defaults.
- "Going dark" is an extreme defensive measure justified only during and shortly after a hacker attack. It amounts to preemptive blackmail.
- The FCC and Congress have done basically nothing to halt toll fraud. Prosecutorial statutes on the books at present are insufficient to deter or punish offenders.
- Call accounting and additional authentication are the main add-on tools we have.
- Vendors and NATA provide information and antifraud consulting services.
- All three major carriers provide toll-fraud detection and prevention services.
- Employee abuse can be stopped using the same call-accounting software. There is a managerial question as to whether it causes more harm than good.
- A toll-fraud prevention program comprises all the above elements.

Notes

1 John Haugh, *Toll Fraud and Telabuse: A Multibillion Dollar National Problem.*

2 "Remote Access Toll Fraud: Detection and Protection," available from NATA, Dept. 5137, Washington, D.C. 20061-5137; telephone: (800) 538-6282, Ext. 260; Price: $53; $33 for NATA members. Also contact the National Fraud Information Center at (800) 876-7060; the Federal Trade Commission at (202) 326-2402; and your local Better Business Bureau.

3 Certain call-accounting vendors' wares have been certified as acceptable under the NetPROTECT plan. For example, Securicom's [Valenic, Calif.; (805) 294-9121] Fraudfinder family of products have met compatibility standards for AT&T's NetPROTECT Premium Service toll-fraud monitoring services. These products include Fraudfinder, a disconnect system; Phoneblock, a trunk lock system; Portguard, a barrier system for maintenance ports; and Fraudfinder 500, a debit card system for DISA and access to long-haul carrier services.

For further reference

Infosecurity News Magazine is free to qualified subscribers. Call (508) 879-9792.

"Stop Toll Fraud Now!," a free 10-page booklet from AT&T. Published in 1994, ask for number PM-7008.

"Toll Fraud, a Tactical Guide for Telecommunications Managers," a free 38-page booklet, also from AT&T, is a reproduction of pertinent parts of John Haugh's book noted above. Published in August, 1993; ask for NET TEXT-001.

"Security Handbook" from AT&T is available free to AT&T service agreement customers. If you do not have a service agreement, it's $65 from the AT&T Customer Information Center at (800) 432-6600. Order Document No. 555-025-600. An Individual Learning Program is also offered using this handbook, a video, and a workbook. Price: $125. Order number 555-025-601. This is also free to those having a service agreement.

Request a free copy of *Access Electronic News* from AT&T if you have an AT&T business communications system such as a PBX or voice messaging system.

Be sure to ask your call-accounting vendor if they are qualified for your carrier's loss protection program.

Chapter

12

How Much Security Is Enough?

Objectives

At the end of this chapter you will be able to:

- Place security in its proper place in your organization
- Apply the five principles that identify needs for protection
- Relate these five principles to one another
- Incorporate additional considerations into security planning
- Distinguish between baseline and extended protection methods
- Discuss risk assessment methods, both informal and formal
- List the advantages and disadvantages of each
- Explain how to get management support for security needs

How Much Security Is Enough?

Perhaps the best way to gauge one's security needs is to place some bounds at either end. The lower boundary is no security at all—everything is wide open to everyone. The benefits are no administration, no expense, and no employee productivity loss. Security is at the very bottom of the totem pole in terms of priority. Productivity will be heightened by free and open access to whatever information helps get the job done.

At the opposite bound, we have security as the overriding, absolute concern, taking priority over any given application or organizational activity. Huge sums are spent on security at the expense of other things, including applications that target the organization's mission. The administrative costs are breathtaking.

The restrictions are so multilayered and so severe that a large portion of employee time is spent on security overhead—logging in, over and over again through fat security layers. File and application access is severely restricted to the point where the system seems to be more of a nuisance than an assistance.

The system can become so oppressive that people begin to find ways around it. They keep stuff in their drawers and pass it hand to hand. They write down passwords. They make diskettes and bypass system security. *Ultimately, oppressive security is self-defeating.*

Each of us in our own systems, and even in our different applications and parts thereof, needs to set our security point somewhere between these two extremes. Some will tilt toward the looser end, while others need tighter security. That equilibrium is what we refer to as *security balance,* just the right amount of security for your organization, in your context, to secure your organizational mission, taking into account your organization's culture and way of doing things (see Fig. 12.1).

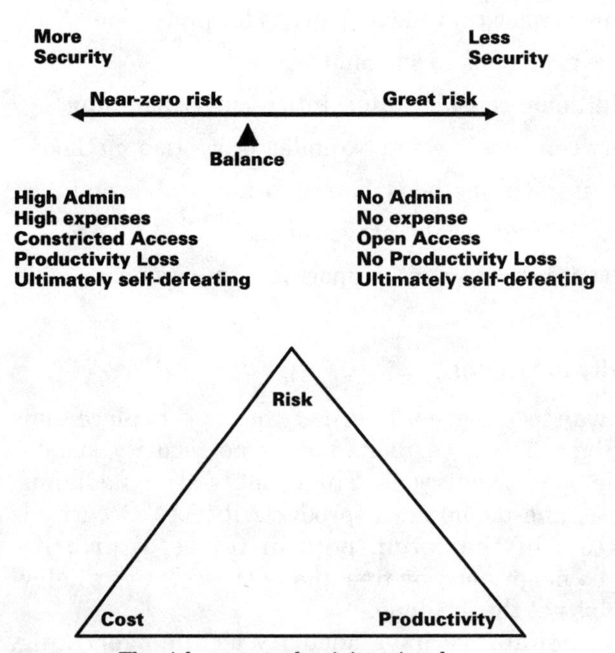

Figure 12.1 The risk–cost–productivity triangle.

TABLE 12.1 How to Assess Security Needs

✓ ***Identify assets***

Inventory networks, network components, locations of information. Remember, data themselves are an asset.

Prioritize networks in descending order, from most vital to least.

✓ ***Identify threats***

Itemize threats to the assets you have listed.

Assets not threatened need minimal protection.

There is rarely a 1:1 relationship between threats and assets.

✓ ***Identify vulnerabilities***

A vulnerability is the path a threat takes to get to an asset.

Protective measures block such paths.

Creative thinking and *brainstorming* can find vulnerability paths.

✓ ***Consider the risks***

A risk is the probability of a threat getting to an asset.

It is usually stated over a period of time such as a mean time between failure (MTBF).

✓ ***Take protective measures***

Protective measures block vulnerability paths and so reduce risk.

Multiple measures may cumulatively improve or degrade overall security.

There may be cost, legal, or moral limits to protective measures that can be taken.

It may be less costly to fix the problem afterward than to prevent it.

Sometimes preemptive extended security measures are implemented because they are easy or inexpensive to do at the time.

Evaluating the efficacy of a protective measure is difficult. If the loss never occurs, did your protection block the threat or was it never there in the first place?

Finding this balance is a little bit of art and science. This chapter presents five basic principles that will help you arrive at a proper balance for your organization (see Table 12.1 for an outline).

Identify Assets

The essence of security is the protection of assets. Therefore, the first order of business is simply to list those assets requiring protection. This might take the form of individual network components and the data they carry. It could well include more than just the network: It might include the building in which the networks reside as well as the people who make the asset useful. The scope is up to you, but our focus here is on networks.

The second step is to classify those network assets into an order of priority. The most vital networks and component assets thereof are those without which the organization cannot continue to function. If assets of this order are not preserved, then all else is meaningless.

Identify Threats to the Assets

If an asset is not threatened, then it does not need protection. Contrarily, a threatened asset and one that is essential to survival clearly needs protection. These threats may be from people such as hackers and employee telephone abusers or from users' unintentional actions that harm an asset anyway. An example would be accidental erasure of a file or disk or an inadvertent sequence of operations that causes damage to an asset. Threats include natural causes that could damage assets, such as fire, flood, or even a massive telephone company central office failure.

There is an implication that threats and assets correspond one to one. This is rarely the case. There are often several different threats to a given asset. For instance, information may become outdated, may be corrupted, or may simply be erased. Multiple assets may be threatened by the same thing, such as a building fire.

It may be that an asset whose loss would be of little consequence is well protected simply by the existing circumstances. The purpose of identifying threats is to see just how vulnerable you are.

Identify Vulnerabilities

A vulnerability is the path a threat takes to get to and affect an asset. Security measures block the path from a threat to an asset. For instance, port protection on a maintenance port blocks the path between a hacker and the PBX and so removes the vulnerability of the maintenance port.

One problem with vulnerabilities is that they can be hard to identify. While some are obvious, the number of possible paths to the asset is usually very high and the distinction of paths consequently obscure. Many of these paths become blocked at some point for some simple reason. The objective is to identify those paths that are not blocked. In many ways, vulnerability identification is a needle-in-the-haystack problem (see Fig. 12.2).

Networks are physically vulnerable to tapping, alteration, jamming, equipment tampering or failure, or circuit failure.

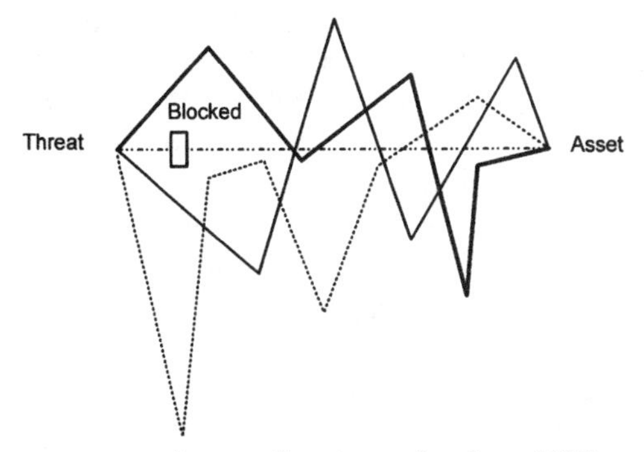

Figure 12.2 Assets, threats, and vulnerabilities. Each line represents a way for a threat to get to an asset. The lines are often indirect. Protective measures block one or more vulnerability paths. Vulnerability paths can and do interact. Very obscure vulnerability paths have the effect of blocks. Paths can branch, with varying probabilities

Vulnerability analysis is dampened by a lack of creative thinking, the cure for which is *brainstorming,* discussed shortly.

Consider the Risks

Simply put, a risk is the probability of a threat getting to an asset, generally stated within a period of time: examples include mean time between failure and the number of traffic fatalities per year. Sometimes risk analysts will try to place an expected dollar loss on a risk. This becomes hard to do when the risk is not quantifiable, as is the risk to a human life.

Take Protective Measures

Protective measures close vulnerability paths or at least make them harder to take. They consequently reduce the ever-present risk. The key question here is how much the protective measure is worth relative to the asset. We do not lock paper bags in a vault (unless they are filled with money!).

Sometimes multiple protective measures collectively multiply security or reduce it. They are not necessarily complementary or synergistic.

Sometimes security measures cost significant money or staff time to carry out, as with multiple logins. There may be legal or moral limits, as in eavesdropping on employees' electronic-mail or telephone conversations. At some point, the cost of implementing the measure outweighs the damage that might be done if the measure is not implemented. So we take a calculated risk that this event will not happen, and if it does, we will activate a security response. As noted before, sometimes it is cheaper to fix the problem than to prevent it. Protective measures are occasionally implemented because they are easy to do at the time. For instance, installing a LAN hub with scrambling security is easy to do at installation time and somewhat harder to do after the LAN is up and running.

The efficacy of a protective measure may be hard to evaluate. The deterrent effect of stiff company security policies may be hard to gauge. The threat of prosecution and hard time in jail clearly does not deter phone hackers, who between 1991 and 1993 hit 550 International Communications Association members for a total of $73.5 million and have now spread to attacking smaller, less well-protected organizations.

Implementation of encryption, for example, is hard to evaluate for effectiveness. It may have stopped someone from bothering to tap a line or stymied someone who did. Perhaps no one ever thought of doing such a thing. You will never know.

Security Responses

In the case of a data network, the overall risk is fairly high. By their nature, data networks are far flung and often beyond the user's control. They are affected by backhoes tearing up fiber cable in Kansas or by a microwave tower that topples in Illinois.

Given our inability to control all network events, it is incumbent upon us to have a backup plan in place. If the line is lost, some local processing should be possible. This is typical of many banks that can do offline processing for a while if the line goes down to the host computer or if the machine itself fails.

Tying the Principles Together

We have here a series of relationships (see Fig. 12.3):

Assets oppose *threats*.

Protective measures oppose *vulnerabilities*.

Risk can never be zero, so *responsive action* will eventually be required.

We now discuss further security considerations (see Table 12.2).

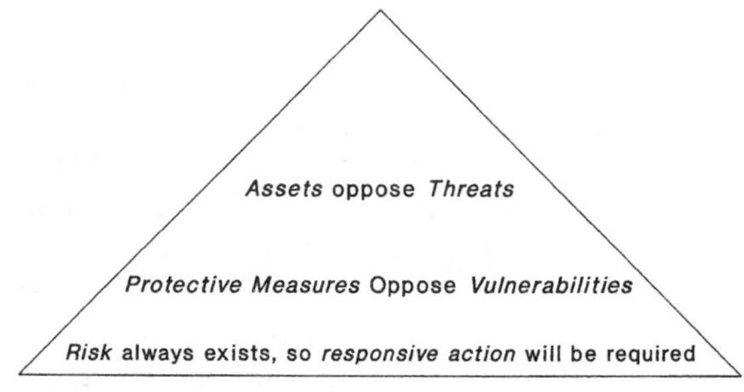

Figure 12.3 Tying the principles together.

TABLE 12.2 Added Considerations

✓ Does the information have *time value?*

Information with little time value is, in a way, self-protecting.

Good: Time to acquire > Information's useful life

Bad: Time to acquire ≤ Information's useful life

✓ Consider your expense-to-difficulty ratio

Measures that can be implemented inexpensively yet give hackers major headaches are worth doing.

Good: Their difficulty > Your expense

Bad: Their difficulty ≤ Your expense

✓ Consider the attacker's perceived cost-to-benefit ratio

Make an attacker's life so miserable that they will go out and find honest work.

If an attacker perceives a greater benefit after much cost in time or money terms, then we need stronger protection methods to tilt the equation the other way.

Good: Cost>Benefit

Bad: Cost<Benefit

Some hackers have six-figure incomes, so our protection needs to be stronger.

✓ Differentiate baseline versus extended protective measures

At its simplest, baseline protection is included in the product at the time of purchase.

Extended protective measures go beyond that. An example would be antivirus software for an NOS, whose baseline protection includes passwords and the like.

Are extended measures worth the reduction in risk? Sometimes extended measures are so inexpensive that we do them regardless of the risk reduction.

Time Value of Information

In considering the degree of protection needed for your voice and data, think about its time value in security terms.

Stock quotation data change moment by moment. It is essential that they be continuously correct. Especially, quote data must be delivered as quickly as possible. Encrypting such data is a waste of money. The time required to crack the code, likely years, far exceeds the time value of the information, which is gone in a flash. Simply authenticating such data is sufficient because a hacker could make a broker sell or buy if the hacker replaced one quote with another.

Conversely, diplomatic embassies around the world encrypt their radio and cable communications with their respective countries. Outside knowledge could influence treaty making and even shift the balance of world power.

Such was the case with the Zimmermann telegram, in which Germany promised Mexico territory in Texas, New Mexico, and Arizona if Mexico declared war on the United States at the start of World War I. The balance of power was again tipped when the United States intercepted Japanese signals indicating Japan's plans for what proved to be the Battle of Midway.

Expense-to-Difficulty Ratio

Broadly speaking, the more we spend on security the more difficult we make it for an adversary to read or tamper with our communications. Ideally we make it difficult enough so that no one will try. If they do elect to try, we want to make it so expensive and so time-consuming that it becomes cheaper to do whatever they want to do honestly.

Thus there is a strong undercurrent of deterrence. It becomes a war of attrition when we make it so difficult for hackers to crack our DISA port that they find the cost-to-benefit ratio unfavorable. That is our objective.

We have to keep an eye on what is happening in the security technology marketplace. With time, DES will become breakable in a time period in which cracking it might be worthwhile and economic. When that time comes, its deterrent power will be lost. We must have a replacement online before that time or else we risk exposure.

Sovereign nations have determined cryptanalysts and basically unlimited funds and resources. Increasingly those resources are being turned against us economically, not militarily. We are the new targets of a threat on a scale we have never before experienced. These are hackers with a difference: they are professionals.

In practical terms, you must look at information from the viewpoint of a hacker. What could *you* do with the information if you had it? Could you wind up with a net profit after deducting the cost of obtaining it?

Baseline versus Extended Protection Measures

One distinction that must be made is between *baseline* and *extended protection.* Baseline protection comes with your NOS in the form of passwords, directory rights, and the like. Passwords in applications, buried login scripts, intruder detection, and so on all add a measure of baseline security inherent in the systems you have. All of these should be used. Rudimentary things like physical access security and basic security policies are baseline measures that are the result of simple prudence.

The next step is the use of extended protection measures for those assets that demand it. This might include encryption, authentication, bridge firewalls, and the like. The question you must ask is whether baseline security is sufficient for each asset you identified earlier.

Risk Analysis

There are two ways of evaluating the probability of an asset being lost because of a vulnerability to a particular threat. The first is qualitative and the second quantitative (see Table 12.3).

The qualitative method is intuitive, "going with our gut." We do it daily when we decide when to cross the street or whether to sneak out of the office early. Baseball players constantly analyze the odds of stealing a base or two.

We are so accustomed to assessing risk intuitively that we have come to trust our intuition. For the most part it has been reliable.

The problem is that intuition can be way off. Just looking at a 500,000-pound Boeing 747 tells you that it could never fly.

In a group of 23 people, the odds are *better than 50-50* that two people in the room will have the same birthday. Intuitively, that seems unlikely, but probability analysis proves it so. So did aerodynamic analysis on the airframe that became the 747.

The more technical reader will hasten to point out that empirical testing of the feasibility of the 747 was essential to its coming into

TABLE 12.3 Ways to Evaluate Risk

✓ Qualitative intuition
- Everyone does it
- Generally cost effective
- Can be done spontaneously
- Requires no data or software
- Accuracy is always suspect

✓ Quantitative risk analysis
- Cost effective?
- Accuracy of data: GIGO
- Skills are needed
- Software
- Rigorous
- Time consuming
- May contradict intuition

✓ Exposure analysis
- Multiplies exposure by vulnerability
- The result indicates the greatest threat sources
- Simpler than risk analysis
- Avoids need to consider nature of exposure

✓ Scenario analysis
- Starts like this: "What if..."
- Brainstorming
- The Tiger Team approach
- Threat logic can be included to add discipline
- The Delphi technique is more focused than brainstorming

✓ Ask
- Ask those who know the system best: the users
- Surveys and questionnaires
- Many know of breaches, but need to be asked

✓ Checklists
- The shotgun approach
- Good as a quick response
- Effectiveness versus cost is always a question

being. That is true in security, too. Testing the system is just as essential if you really want it to work.

Quantitative methods

Risk analysis can also be performed quantitatively. The concept is simple. Let us say that we have a network whose usage generates $200,000 in revenue per 20-day month. Let us also say the network has a mean time between failure of one day every three months. From these figures we calculate a monthly loss expectancy:

$$\text{Monthly loss expectancy} = 0.33 \text{ day per month} \times \$10{,}000$$

$$\text{Monthly loss expectancy} = \$3{,}300$$

If we examine the network and find that the main source of trouble is wiring (as it is nearly 70 percent of the time), then a wiring hub might be a good idea instead of the rat's nest we have now. The hub will then wipe out 70 percent of the loss or $2,400 per month. If we spend $7,000 on the hub, we can recoup our investment in three months and be ahead $2,400 per month from then on.

Risk analysis software

Risk analysis is implemented in much more sophisticated forms using spreadsheet software packages that run on a variety of platforms. Prices range from $100 to $17,000 or so per license. Depending on the products, they may be more or less sophisticated and detailed in their analysis of your system and estimating the degree of risk. Some even use expert systems to gauge the probability of particular risks.

NIST offers the "Guide for Selecting Automatic Risk Analysis Tools,"[1] Publication No. 500-174 which is available for a nominal fee. There is also FIPS Publication No. 31[2] which explains the use of risk analysis. FIPS Publication 65[3] is a how-to guide to performing risk analysis. Sometimes such an analysis is required of government contractors' security programs.

Cost effectiveness

It is unwise to conduct an expensive risk analysis before obtaining general management backing for a security review. It is better to start with a quick analysis over a few days using just a few people. Then it is wise to present initial findings to see if further work is needed and where. Some forms of attack and protection are plainly evident, and including them in a risk analysis is a waste of time and energy.

Data accuracy

You have no doubt heard the computer litany, GIGO. It stands for garbage in—garbage out. It is nowhere more true than in risk analysis. Risk analysis databases contain probabilities for all sorts of things—fires, earthquakes, and so on. Some of these data change regionally, so be sure that the data are relevant to your part of the country.

Exposure Analysis

If you are allergic to major-league number crunching, consider *exposure analysis*. Exposure analysis considers the degree of exposure, regardless of the source (people, nature, accidents, etc.). We assign that degree of exposure a value, and then multiply it by a weighting number based on vulnerability: the greater the product of the two, the greater the need for protection.

Scenario Analysis

This involves imagining a vulnerability path that leads a threat to an asset. The likelihood of a change in direction by event can be evaluated statistically or by your own judgment, with a consequence at the end. Often, a combination of events must occur to cause a loss of an asset. This is the logical AND condition. If there are several combinations of events that can cause the loss, this is the logical OR condition.

Therefore it is possible to model scenarios using "threat logic." Similarly, security faults can be combined through AND and OR operations together to lead to a security failure. Originated by security consultant David Snow, threat logic is a rigorous way to examine scenarios that could cause loss.

It is a lot more fun to do *brainstorming*. A bunch of people sit around a table and dream up the most exotic (or simplest) ways to attack a system. Vulnerabilities are identified that might otherwise go unnoticed.

A final method is the *Delphi technique,* originated by the Rand Corporation think tank. Take a number of network people and ask them for vulnerability estimates by numerical weight. The accumulated numbers are averaged and the estimators asked to consider the averaged numbers in reevaluating their estimates. A more realistic consensus is the usual result.

Ask!

One of the best ways to analyze risk is to ask those most deeply involved. Questionnaires are especially useful in determining the location of computer workstations, connections needed to the network, and uses of the network. Sometimes users will discover a security breach by accident but never tell anyone—*because no one ever asked.*

Checklists

After much experience, a pattern emerges as to what constitutes the most common vulnerabilities and useful protections. Rather than use many exotic and time-consuming analyses, just use a checklist.

The problem with a checklist is that it is a shotgun approach. It has to be applied within your context, not by rote. It does not account for your particular vulnerabilities or the protection dollars involved. That is up to your judgment again.

Some vendors offer a checklist running into nearly 1,000 items. Presumably if all were implemented you would achieve the equivalent of Fort Knox security. More often, the entire list is so intimidating that people throw it in a corner and forget it. This amounts to zero security improvement.

True to form for a security book, this book contains a series of checklists in Chapter 14. We have tried to be realistic and focused in creating them. They *are* a good short-term "Let's *do* something!" response, but are not substitutes for some good solid analysis on the part of you and your fellow staff members.

Gaining Management Support

Having recognized the need for improved security, enlist management's support (see Table 12.4). Conduct a brief analysis—weaknesses should have their possible annual losses noted.

Remember: you are *not* building a case. You are making an *assessment.* The difference is *objectivity.* You are not conducting a needs analysis if you have already decided that additional security measures are or are not warranted. You are instead looking for facts to support your conclusions. Management is well aware of this tactic—they use it all the time.

Schedule a two-hour meeting (tops) with upper management. Begin with a training video or a summary of losses by firms similar to yours. Present any circumstances that may have heightened your interest in security as a concern: perhaps an incident that threatened the organization. Make your presentation relevant or you will lose your audience fast.

Use high-quality overheads. Keep it simple. Always define a term or acronym the first time you use it; better still, avoid them. You are making a business presentation, not a technical one.

TABLE 12.4 How to Make an Effective Security Presentation to Management

- ✓ Conduct a Brief Analysis
  - Be objective
- ✓ Schedule a meeting (two hours at the most)
- ✓ Introduce security as the topic
  - Training video
  - Losses in similar organizations
  - An incident that prompted the brief analysis
  - Relevance is vital
- ✓ Keep it simple
  - Use professional overheads
  - Avoid technical jargon
  - Make a business case
  - Consider a joint presentation
- ✓ Summarize the result of your brief analysis
  - Indicate vulnerabilities found
  - Quantify into potential dollars or assets lost
- ✓ Propose a solution and a plan of action
  - Include all expenses, direct and indirect
  - Present security as a prudent fiduciary measure
- ✓ Ask for approval
  - Leave them with additional material
  - If not approved, find out why, fix it, and try again
  - Sometimes it is just the wrong time
- ✓ Weak budget support for security is common although there is lip service to the contrary
- ✓ Failure to present a business case
  - Often the reason for weak budget support
- ✓ Do not give up

Consider a joint presentation with someone else. It adds credibility and a bit of variety.

Summarize the result of your brief needs analysis. Indicate deficient areas that need further study and beefing up. Again, present the problems objectively, analytically, and in a businesslike way.

Having defined the problem, propose a solution. You may wish to form a security working group, consisting of representatives from several departments. This is a very good idea, because they have a different perspective and will have to live with whatever is implemented.

Set forth a plan of action, including a timetable and evaluation milestones for management oversight. Management needs a way to measure progress and detests being left out of anything.

Staff time is a precious resource these days. Estimate the person-hours involved and any additional expenses such as software or consulting services. Do not try to sugar-coat your proposal.

Remind management gently that it has a fiduciary responsibility to the shareholders and employees to protect company assets. The security project is a prudent investment that no shareholder would likely criticize.

It is unlikely that you will receive an instant yea or nea. Leave the managers with some articles or additional reading that will give them more background. Most dislike making decisions with insufficient information, and some need more than others to feel comfortable.

If you gain approval, carry out your plan as you were mandated to do. If you are turned down, find out why, fix the problem, and try again when the time is right. Sometimes it is just a matter of timing. Do not give up if your needs analysis showed weaknesses. In the meantime do what you can within your means and budget to lock things up.

Many top executives say security is either "extremely important" or "very important" to their organizations, yet fewer than 6 percent of respondents in a study said they dedicate more than 5 percent of their network budget to security. Basically, they pay it lip service.

The failure is not only with them. It lies also with us who have failed to make an adequate case. If you follow the suggestions above, you will stand the best chance of doing that.

In the next chapter, we discuss how to establish a comprehensive security policy.

Review

- Security can range from extremely tight to nonexistent. At worst, it crimps activities, and at least, it exposes the organization to great risk.
- A balance is needed between cost, productivity, and risk.
- An assessment consists of identifying assets, threats, and vulnerabilities.
- The risks are then considered and protective measures taken.

- Assets oppose threats; protective measures oppose vulnerabilities.
- Since risk is never zero, a restorative measure will be needed one day.
- Information has time value and must be considered in its protection.
- A small expenditure that increases an adversary's difficulty is good.
- The object is to make it unprofitable for an adversary to attack you.
- Risk analysis may be intuitive, informal, and qualitative.
- It may also be formal and quantitative.
- Exposure analysis and scenario analysis are two more ways to assess risk.
- One of the best and simplest ways is to simply poll your users.
- A quick-response risk aversion method is to use predefined checklists.
- Management support tends to be more often vocal than monetary.
- To get real support, do some research, a quick needs analysis.
- Present your findings professionally and objectively.
- You are making a business, not a technical case.
- If at first you do not succeed, do not give up.

Notes

1 "Guide for Selecting Automatic Risk Analysis Tools," NIST Publication No. 500-174.

2 "Guidelines for Automatic Data Processing: Physical Security and Risk Management," FIPS Publication No. 31

3 "Guidelines for Automatic Data Processing: Risk Analysis," FIPS Publication No. 65

Chapter

13

How to Create a Network Security Policy

Objectives

At the end of this chapter you will be able to:

- Explain why network security plans fail and how to avoid failure
- Formulate an effective network security plan
- Identify the information needed in a quick look and its sources
- Compose an effective security questionnaire
- Outline elements of a security task force charter
- List areas to be scrutinized
- Design a workable implementation plan
- Know when security through obscurity is an appropriate policy
- Recount the inviolate rule of security

Network Security Policy

Network assets worth protecting demand a professional, organized response. That is why a network security policy is mandatory. It shows that the issue has been thought out and committed to paper. It avoids potential claims of arbitrary enforcement when the imposition of disciplinary action becomes necessary. It tells you what to do when an incident occurs. In short, a network security policy shows management that you have done your homework.

A security policy is a reference document, but it is not static. It is subject to amendment as systems and circumstances change. It should be reviewed at least annually, and more often than that if your environment changes rapidly.

From the outset, it is essential to acknowledge that executives, managers, and individual users all have security responsibilities. These responsibilities are summarized in the Tables 13.1 to 13.3.

Acceptable Use Policy

It is not reasonable or fair to hold people accountable for nebulous or unstated security rules. An acceptable use policy makes clear what standards will be enforced. At a minimum, an acceptable use policy should include:

- Password management, protection, and change rules

TABLE 13.1 Executive Responsibilities*

- Set the security policy of the organization (nothing short of global security is sufficient).
- Allocate staff, funding, and positive incentives.
- State the value of information to your organization.
- State your organization's commitment to its protection.
- Make it clear that this responsibility is everyone's job.
- The above said, assign responsibility to specific individuals.
- Require computer security and awareness training.
- Hold employees personally accountable for the resources in their care.
- Monitor and assess security through external and internal audits.
- State penalties for nonadherence.
- Goals include:

  Risk reduction to an acceptable level.

  Assure operational continuity.

  Comply with applicable laws and regulatory requirements.

  Assure the integrity and confidentiality of information in your hands.
- As with every dimension of leadership, *you* will set the example.

*This is a brief abstract of NIST's FIPS Publication No. 500-169.

TABLE 13.2 Management Responsibilities*

- Assess the consequences of a security breach in the areas for which you are responsible. This includes outright network failure due to a virus or other cause, corruption, or theft of information under your care. Risks include:

  Inability or impairment to perform necessary tasks.

  Waste, misuse, or theft of funds or resources.

  Internal and external loss of credibility.
- Be cost effective: find the right balance of acceptable risk versus dollars spent on security.
- Assess operational continuity: How long can each system you have be down before the consequences become unacceptable?
- Maintain accuracy of the data in the system.
- Maintain the necessary degree of confidentiality.
- Comply with any laws or regulatory agency directives.
- Set an acceptable use policy.

*This is a brief abstract of NIST's FIPS Publication No. 500-171.

- Strictures prohibiting the illegal duplication of software
- Antivirus policy with specific do's and do nots (see Chapter 2)
- Encouragement to use screen savers and automatic log out features
- Rules regarding the use of organizational computing resources for personal use (you may wish to be a little liberal in this regard; this is an excellent way for people to learn).
- Penalties for violations; an annual review item; other human resources concerns.

Why network security plans fail

If it is to be effective, a network security policy cannot be implemented by decree. Since it affects every user of the system, its reach goes far beyond the information system department's walls. A policy, no matter how sound, that is not established by a consensus of those to be governed by it is a sure way to fail. Some of these reasons for failure are shown in Table 13.4.

Cummins Engine Company in Columbus, Ohio has an Information Security Council consisting of volunteers from the computer, telecommunications, legal, auditing, marketing, and

TABLE 13.3 User Responsibilities*

- You are individually responsible for protecting the data and information in your hands. *Security is everyone's responsibility.*
- Recognize what data are sensitive. If you do not know or are not sure, *ask.*
- Even though you cannot touch it, information is an asset, sometimes a priceless asset.
- Use the resources at your disposal only for the benefit of the organization that provided them. Do personal work on your own system. Do not use computing resources for anything illegal.
- Abide by the acceptable use policy provided by your organization.
- Understand that *you* are accountable for what *you* do on the system.
- If you observe anything unusual, *tell your supervisor.*
- Some specifics:

  Do not share or disclose your password.

  Choose a password that would be hard to guess.

  Log off before you leave your terminal.

  Lock up sensitive information kept on disk or paper.

  Do not give others the opportunity to look over your shoulder if you are working on something sensitive.

  If someone in your area looks like they do not belong there, *speak up.* Ask them their business in your area.

  Protect equipment from theft or damage due to water, electrical surges, and the like; keep it away from cigarette smoke, food, and drinks. Know where the fire extinguishers are.

  Do backups religiously.

  Do not do things that would permit a virus to spread.

*This is a brief abstract of NIST's FIPS Publication No. 500-171.

human resources divisions. Such teamwork helped tighten mainframe security, solve an unauthorized software duplication problem, and set up an Internet application firewall. The firewall effort required telecommunications, networking, and software people to work together. In other words, it took teamwork.

At Northwest Airlines Corp., the information security division learns from the human resources department when layoffs and voluntary departures occur. Conversely, the human resources department needs assurance that this advance information will be kept confidential. All understand that disgruntled ex-employees can be a real security threat.

TABLE 13.4 Why Network Security Policies Fail

1. The policy is implemented in a vacuum or by decree.
2. It is not coordinated with other policies and the disaster recovery plan.
3. The plan or part thereof is illegal.
4. The focus is misplaced on technology and not on people.
5. The policy is not promoted or made sufficiently visible.
6. "It cannot happen here."
7. A lack of penalties makes the policy hollow and toothless.
8. A predestined sense of failure exists, e.g., "It is impossible to get our people to do this."
9. There is a failure or an inability to see the need for security policy.
10. Users may fear that they will be locked out of useful resources.
11. A lack of focus exists, exemplified by an excessively broad or too narrow a scope (e.g., people, systems, applications).
12. Lukewarm management support is given. A survey by Ernst and Young LLP found that 42 percent of senior management in U.S. companies considered information security to be "somewhat" or "not" important. Fully 15 percent devoted no full-time resources to security, even though more than half the respondents reported some kind of information loss. For those citing "mission-critical" applications on their LANs, half said the attendant security was "unsatisfactory."

Northwest also has a policy regarding the handling of files and electronic mail. Shall the mail be rerouted, returned, or inactivated? Who will inventory the files and see what must be preserved and what deleted?

As you can see, establishing and maintaining a network security policy is an interdepartmental process. It calls upon the network manager to work with other departments toward a common goal.

As above, network security plans will not work in a vacuum. They must be integrated with overall information systems and telecommunications security policies and the disaster recovery plan. (Please do not say you do not have one.) They must be consistent with all company policies including those originating in the human resources department.

Policies must also be legal. In some states it is illegal to use a lie detector to hire an employee and illegal to monitor their telephone conversations, as many help desk and telemarketing managers do where allowed.

Security is always a people issue, not a technology issue. Properly addressed, employees will not only abide by security policy but enforce it on coworkers who stray. Improperly addressed, employees will collude with one another to get around rules they think silly, cumbersome, or arbitrary.

A dusty policy book sitting on a shelf does not do much to improve security. As a people issue, it must be resident in the fore-brain of every network user. This means that a network security awareness campaign must be an ongoing part of network policy.

The federal government is a superb example. You cannot go anywhere in a federal facility without seeing signs on bulletin boards promoting security. Security is firmly implanted in every employee's forebrain.

Some users think security breaches cannot happen to them. They, too, need some attitude adjustment. Computer viruses are not too common, but they do happen. If you are struck by a virus for example, publicize the experience as a way to promote sound antivirus policy. To the same extent, publicize security breaches reported in the news or trade magazines.

Policy violations have to carry a penalty or they will be ignored. The policy should state the penalty for a first infraction, such as a warning, leading up to a negative review and even dismissal. For example, restitution could be required from a person who transports a virus into the network and causes a loss.

Some managers believe the task of establishing and maintaining security is impossible because of today's dynamic distributed computing environment over which they have little control. As decision making migrates to the PC level and employees acquire notebook machines, data are on the move. Not only are data moving targets, the risk of exposure increases as our organizations become less hierarchical. Here is a classic case of misfocus, on technology and not on people. It is people who operate all these systems, wherever they are.

A few managers do not see the need for a security plan. They have difficulty in conceiving the risks. That is why it is necessary to publicize losses due to security breaches. Simply because the odds favor it happening to someone else does not mean that it cannot happen to them. In that case the odds jump to 100 percent. Upper management will want answers if there is damage to the business because of such a laissez-faire attitude.

Users bridle at a security plan that locks them out of areas where they formerly had open access. It is our job to see that security does not improperly impede employees going about their daily tasks. It is also our task to see that they do not go places where they have no need to be. We must find and maintain that balance. It is important to explain to employees that the less access they have, the less exposure they take on in case of a security breach.

Policies also fail due to a lack of focus. Support for security is an apple-pie issue: it is impossible to be against security. Blanket policies inevitably fail. Tailored policies succeed and this is why we conduct an initial assessment.

Building an Effective Security Plan

The converse to the above is embodied in Table 13.5 and Fig. 13.1. It is possible to build an effective security plan, and the balance of this chapter is intended to illustrate how to create one.

Scope

Network security is one element of computer security and an element of overall organizational security. Early on in the process, beginning with the "quick look," decide on the scope of the inquiries to be made.

While it may be easier to deal with a limited scope, the result may be incomplete and inconsistent with the balance of the security structure. The scope of the inquiry and contemplated changes must be broad enough to fit comfortably into the larger picture, but not so broad that progress becomes impossible.

TABLE 13.5 How to Build an Effective Security Plan

- Admit that you have information requiring protection.
- Executives, managers, and users all have their role in security.
- Develop specific plans to protect your information.
- Begin at the outer edges of the information or telecommunications structure.
- Establish standards once.
- Develop consistent standards for all departments.
- Consistency will limit costs and improve productivity.
- If it is protected on the mainframe, it must be protected in a PC.
- All machines must be secured.
- Perform some kind of risk assessment, whether formal or informal.
- Maintain a global, enterprise-wide security philosophy. Global security is more effective, cheaper, and less visible to users.

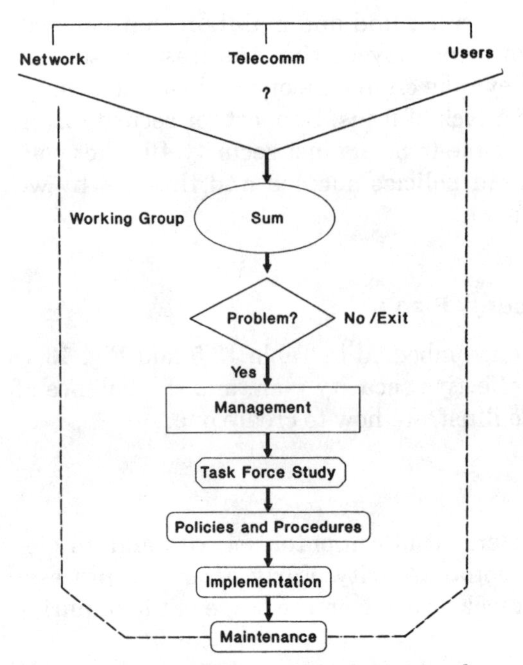

Figure 13.1 Developing the network security plan.

The Quick Look

As we mentioned in the last chapter, a quick look by a small working group is the best start toward a workable, effective network security plan. It is always wise to look before one leaps.

An initial study avoids the trap of implementing preconceived notions as to where security is good or bad. More often than we would care to admit, those preconceived notions turn out to be off or even dead wrong. Approach the inquiry with an open mind. Let it take you where it will.

Perhaps a network security incident triggers the probe, or maybe it is just something everyone knows needs to be done. Perhaps it is in the list of projects for the year. Lacking these catalysts, perhaps you need to take the initiative.

The purpose of a quick look is to find out if there is reason to go further without investing a lot of staff time and resources. If there is no reason to go further, or if the needs prove minimal, the inquiry stops here.

Otherwise, the outcome will be used to modify current policies, create new ones, and/or implement additional security precautions.

Some of these activities involve significant expense, so upper management approval will be needed.

The initial review develops facts and figures that can be presented objectively to management. They carry weight because they are current and relevant. Preliminary findings set the direction for further work if needed.

Sources of Information

Information sources boil down to you and your fellow staffers and the user community. Both are represented in the working group. The great value here is that you have diametrically opposed perspectives on the network as providers and users. Between the two of you, you form a composite whole that approaches your organization's perspective.

The user community

People who live with the system, day in and out, get to know it pretty well. Often they know it better that the providers. (Ask any help desk analyst).

A questionnaire or survey of the user community is very useful. Assembling a questionnaire is a bit of an art form. In addition to the example in Fig. 13.2, here are some suggestions:

- State that you are conducting a network security survey and ask for their help.
- Do not ask for names unless respondents wish to supply them voluntarily.
- Automate the survey: put it on a login screen and save the results into a file. It will make the survey visible, make reducing the data easier, and the cost for the entire survey will be minimal.
- If you survey on paper, send the survey to *department managers* for distribution and completion. Receiving a directive from the boss puts a special spin on the survey.
- Ask people which system(s) they use, but do not be so specific that you can identify the individual.
- Ask the user if they have noticed ways to gain system entry through unexpected paths, privileges in excess of those really needed, unusual events that might imply tampering, viruslike

The Network unit of the MIS Department is conducting a review of our security procedures in an effort to identify areas for improvement. At the same time, we are trying to find solutions that will streamline security procedures and so not interfere with your daily activities.

Would you please take a moment to answer these few questions? Thanks.

I use the following system(s):

LAN e-mail Mainframe Mini Other (specify): ________

WAN Windows PC or clone Macintosh

The three most important system tools that help me do my job are

1. ________________

2. ________________

3. ________________

I think we are vulnerable to:

A computer virus Theft Other: ________

A hacker A major disaster

We need protective measures (specify): ________________

Have you discovered a "back door" into a system that we should fix?

If so, what is it? ________________

Do you feel the existing security is

Too restrictive Too loose Just right

Have you experienced a security problem such as a hacker or a virus? If so, what was it? ________________

If you could implement just *one* security mechanism, what would it be?

____

Your name (optional) ________________

Please return to Mail Stop B-114 or e mail: survey @ unico.com

Thanks for your input. All responses will be individually evaluated.

Figure 13.2 A sample security questionnaire example.

symptoms, laxness in controlling physical access, vulnerable systems, and servers or applications that are insecure yet whose absence could damage the organization.

Ask focused questions, neither too specific and narrow nor so broad that users do not know what form their answers should take. It is also important not to convey a sense of the kind of answer you would like. "Is system security satisfactory?" is a poor question both because the user reads it in the context of the requester, and because a yes-or-no answer is possible. Use even-numbered, not odd, graduated responses. These force the user to take a position.

Check-off items on the survey are fine, but ask for measured, thoughtful responses. Indicate that the survey results will be used in a security evaluation: The time they take to answer will not be wasted. Ordinarily, publishing the result of a survey is the best way to get good response. Yet a security survey may show weaknesses that you would rather not publicize. Instead, thank everyone for their input, indicate the response rate, and assure them that their input was used.

If it is possible to conduct personal interviews, do so. In the interest of time, it may be best to focus on managers' views but all strata of the organization should be sampled.

Some enterprises go so far as to conduct many in-depth interviews and even focus groups. If you can afford the time and expense, that is fine. The responses will be that much better in quality.

The MIS and telecommunications community

Network and computer professionals must also be polled, but the smaller number of individuals involved should make it possible to obtain their major views in one or two meetings. An agenda should be prepared to keep the discussion on track. MIS and telecommunications staff may meet separately to discuss unique issues and jointly where information and telecommunications priorities coincide or clash.

Methodology

The general form of inquiry should be the same as that applied in the last chapter: identifying assets, threats, vulnerabilities, protective measures, and responses if all else fails. This consistently applied framework allows information created in the last chapter to be mapped into a larger evaluation.

Form a working group

A working group consisting of information and telecommunications systems and the user community should reduce the information received into tabular form from the surveys and include brief summaries of other vulnerabilities noted. The working group should reach one of three conclusions:

- Nothing more than minor changes are needed.

- Changes are required but they do not call for an organization-wide effort.
- A full-scale review plus substantial changes are required.

In the last case, a formal presentation is required to management in the format discussed in Chapter 12.

The Network Security Task Force

If approved by management, a task force including every element of the organization should be assembled. Subcommittees may be needed if the organization is very large, with a representative from each of them on the task force committee itself. The subcommittees could be organized by department or by the system type to be studied (e.g., telecommunications; LANs; WANs). For focus, it may be useful to have a charter, for which a sample is shown in Table 13.6.

The task force approach can have dramatic results. When the General Accounting Office issued a scathing report concerning the Resolution Trust Corporation's (RTC) security weaknesses, the report gave RTC staff the ammunition it needed to centralize access management by having each LAN user sign an access control form before gaining access. The form also determined whether the user would have wide area access as well, affixing accountability.

TABLE 13.6 A Sample Task Force Charter

Develop appropriate controlled access to hardware, data, and applications deemed sensitive.
Within the above context, achieve end-user transparency such that user productivity is either unaffected or improved by streamlining security procedures.
Provide flexible, modular, and adaptable security features that will not become obsolete as the organization changes.
Develop enterprise-wide security solutions, not piecemeal ones.
Minimize the need for administrative and training overhead.
Incorporate virus prevention, detection, and recovery mechanisms.
Build in audit trail capability.
Make full use of existing security tools.
Identify where baseline and extended protection measures are needed.
Establish minimum security features that must be included in new software, hardware or computers purchased by us.

Deactivated IDs are found by periodically comparing the mainframe access control list to the LAN ID and master payroll list. A chronic virus problem was solved when a LAN administrator was found to be logging in from an infected PC. RTC now uses antivirus software weekly to check individual disk drives. All of this was made possible once the problems were identified and management gave its support.

Even as the task force meets, you may note a subtle change as security awareness penetrates to the core of the organization, rather than residing on the fringes. Group members begin to internalize the goals of security and by example, it begins to diffuse throughout every department even before a new policy is set.

Areas to be scrutinized

The target areas for the task force or working group should include those itemized in Table 13.7. Each area is discussed individually in the following paragraphs.

Physical security. Is your building like Grand Central Station? Do people wander around whom you have never seen before? Perhaps a badge system is appropriate.

Yet badges are no solution if people do not consider security their personal responsibility. The author has visited many facilities where obtaining a badge was necessary before entry. Once inside, he could then wander around, looking officious all the while, badge in pocket (not displayed). Suitably dressed and businesslike, in 20 years he has yet to be stopped and asked, "Who are you?"

We have already noted that vulnerable points such as servers and hubs should be in restricted areas. Where do you violate this

TABLE 13.7 Areas to be Scrutinized

- Physical security
- Access control
- Secure the medium
- Personnel policies
- Audits
- External media policy
- Backups
- Security agreements
- Training
- Administrative policies

rule? Similarly, cables should not run rampant. Unused ports should be disconnected. Unused cables should be removed.

Access control. We have said that group user IDs can be shared, but individual passwords should not. Make the uniqueness of passwords a policy. Then help people carry it out by helping them pick passwords from a human-readable phonetic list. Do not let them pick excessively easy passwords or ones that have been used already. Establish a policy that sharing or writing down passwords is a security policy violation, with penalties for first and successive offenses.

Foster stale password turnover by having the system remind users that their password is getting old (90 days or more). Some systems warn the user a few times and then force the user to do so before they can continue. Others simply do not allow them in any more without contacting the system administrator. Henceforth, they will be a little more mindful (for a while) when the system first politely asks them to change their password.

As noted in Chapter 8, passwords should always be encrypted. Some systems such as NetWare 3.1X offer the option of *not* encrypting passwords. Remember to automatically log users out after a period of inactivity. Someone else could use their workstation.

Multiple systems create a multitude of user IDs and passwords. In such cases, it may be better to use a front-end security server to allow a single login. NetWare 4.X and Vines' StreetTalk typify directory services that maintain security as well as ease of use.

Secure the medium. Encrypt or authenticate as required. No form of wireless transmission is secure without encryption. Even the leased data circuits you rent from a common carrier may pass through microwave links somewhere along their route. It is essential to have an alternate plan in place should a backhoe dig up one of your backbone network circuits. One element of such a plan is to have two entry–exit points for the building's communications. Another is to split traffic carried between several carriers. One affected customer in the Hinsdale, Ill. central office disaster had backup telephone lines stationed in the false ceiling of the cafeteria. When a fire occurred, his company's staff moved to the cafeteria and resumed business. Another had a number of cellular telephones on hand for just such an emergency.

Personnel policies. Some organizations subject employees in sensitive positions to psychological screening and/or polygraph examination before and/or after hire. This may seem extreme, but database or system administrators can break your company if they are dishonest or have serious personality flaws. It is easier not to hire someone than to terminate them after a serious flaw has been discovered. A disconcerting statistic from the U.S. Department of Health and Human Services study showed that *20 percent of the people who committed computer crimes had prior criminal records.*

Employee terminations must be handled with care. When an employee is terminated, they should be escorted out the door promptly with no further computer access allowed.

This calls for a sense of timing. An employee whose career path is about to be changed should not make the discovery that they suddenly cannot login any more before they are terminated. While acceptable from a security standpoint, the only way to let an employee go more crudely is to fire them by facsimile.

Divide responsibilities. It makes the dishonest person's task harder, but not impossible. Donn Parker, a well-known security analyst, estimates that about half the cases he has investigated have involved collusion. So dividing responsibility is no panacea. Honest employees do not mind and even prefer such division as evidence of their integrity and a reduction in their personal exposure.

Publicizing terminations, disciplinary actions, or prosecutions due to security violations is a psychological deterrent. It is a hardball tactic indeed, but it works. The understandable tendency to hush up an embarrassing security breach only encourages hackers. Silence encourages apathy and neglect by the user community. The notion that "It cannot be a problem if it cannot be seen" is obsolete.

Audits. Much as we dislike snooping on our own people, auditing has shown itself to be a highly useful and worthwhile security technique. Be sure to publicize the fact that auditing of login activity, phone and credit card use, and so on is ongoing.

Simple awareness by supervisors and managers means a lot. A fine employee whose performance suddenly slips for no perceptible reason should trigger a warning. Aldrich Ames, the confessed spy, gave off many obvious warning signs that should have signaled his supervisors of a problem, such as living beyond his means. They were either not noticed or worse ignored, until it was too late.

Something out of the ordinary routine may not mean a security lapse. However, it is an unfortunate fact that most computer crimes have been discovered accidentally. An anonymous hotline or mailbox may be useful for tips, just as the Federal Aviation Administration has a "snitch line" for members of the aviation community to report problems of any kind anonymously. Many urban police departments have a similar facility.

Auditing software can produce reports by exception, send them via electronic mail or even trip a pager. The reports may prove to be crucial evidence in court. To be able to tie events in chronological order, system clocks should be set to official time (e.g., the CBS radio network chime on the hour) and reset periodically. Many network security managers recommend that the audit logs be reviewed daily. Though tedious, it does give the manager a clear picture of what is normal and sensitizes them to abnormalities when they occur.

External media. Disks, diskettes, and tapes should all be scanned for viruses before use. Shareware and executable program downloads from bulletin boards should be prohibited into networked machines.

Using these programs and diskettes brought into work from home is not necessarily a crime. An excessively rigid policy barring their use precludes the use of valuable software (i.e., the *Network World* BBS utilities mentioned in Chapter 8). A too-tough policy will be ignored or spited. Simply scan incoming material thoroughly before its use is allowed. Similar rules apply to notebooks connected to a LAN temporarily. A notebook's hard drive should be scanned before connection.

As a corollary, it may be desirable to disallow by policy the transfer of information on diskette from one network to another as a way of preventing a handheld network leak. It is important to let people know why this policy is effected in order to avoid the appearance of being arbitrary.

Backups. Backups should be performed more often if the data change more often. Several backups into the past should be kept. If a file is damaged or lost, one can go far enough back into the past to recover it. It may be necessary to go back several generations to find a backup set that is reliable.

There are many forms of backup rotation plans. Typically the rotation plan includes an older off-site backup and more recent on-site backups. Be sure to keep all backups accessible but under lock and key.

Security agreements. Every employee should receive a summary copy of the company's security policy. Employees should be required to sign a statement indicating that they will comply with the policy. It should be renewed annually, perhaps at the time of the employee's performance review. Special agreements may be needed depending on the activity in which the employee will be engaged.

Training. Training makes people aware of the security rules and helps them understand the rules and their purpose. Proper training leads to acceptance of security rules and is also a deterrent to improper behavior. "Proper" means that the training curriculum is matched to the audience. The curricula will change depending on the system(s) used by a group and their levels in the organization. Production workers need one kind of training; accountants need another.

Administrative tools. Organizational policy should require people to take vacation every year. One habit pattern of abusers is that they dislike, sometimes intensely, the suggestion that they take their vacation.

The reasons are twofold. One is that someone taking over their tasks will see what they have done. Another is that they will lose ill-gotten income during their absence. Some individuals who do not take vacations lead unidimensional lives and this is sometimes read properly as a sign of instability. Insisting that such people take their vacation moves them toward being at least two-dimensional and so toward greater stability.

The Product: Policies and Procedures

Philosophies beget policies that take real form as procedures. If a person uses an eight-letter password as a mandated procedure, then they are implementing a policy whose philosophy is that long passwords are safer. Procedures eliminate ambiguity, provide direction, and are measurable as to accomplishment or not.

The task force may classify information resources as to criticality and mandate backups, redundancy, or encryption as a matter of policy for systems at a given level. Such a policy encourages that appropriate security measures will be applied even after the task force disbands.

Wherever possible, make use of existing policies. There is little point in reinventing the wheel. Policy changes also require a change in behavior. Users find this unsettling and disruptive. Keep it simple.

Implementing Policies and Procedures

The hardest part of a network security policy is its execution. This is not so much the addition of encryption to a WAN, but much more a matter of changing peoples' behavior.

Old habits die hard. Replacing them with new habits takes time, skill, and patience. Often, a year will pass before the new policies really take hold. It takes persistence by the network manager and indeed all of management. This includes the folks at the top. Start your implementation with them in a half-day or so briefing. An announcement of the new policy by the boss is a way of making the new policy official and visible.

Since the help desk has so much user contact, they should be briefed next. Spend a day or even two (not consecutive) days telling them about the security provisions you have made in operating systems, NOSs, audit tools, and antivirus software, with an emphasis on how it will affect their operating environment.

Changes in procedures and policies should be discussed to let them know that there has been a shift. The help desk people set the example for users, so they must understand the policies and procedures thoroughly. A full copy of the network security plan should be in the help desk area, as well as a summary for quick reference. Users should also receive a half-day briefing as to policies and how their systems and work procedures will be affected. It is appropriate and convenient to have all employees sign their security agreements at the end of their respective sessions.

"Security Through Obscurity"

Does it still have a place? The quasiexotic nature of what we do as communications people does indeed provide a measure of security. Some adversaries are deterred by the ever-growing complexity of the communications environment, but not all. A small, intelligent number of hackers have developed techniques as exotic as our own systems. In addition, they have a strong profit motive, whereas we are trying to keep what we have earned. We are in a defensive posture while they are on the offensive.

It is no longer adequate to assume that your dial-in ports are safe because they are lost among millions of telephone numbers. Attack dialers obsoleted the needle-in-the-haystack approach. Nor is it sufficient to presume that your IPX/SPX or TCP/IP LAN protocol will save you. Analyzers can chop packets into byte-size slices any way the hacker wants them.

Reliance on being a needle in a haystack as a threat deterrent grows less reliable every day. If you insist that you are too small an organization for hackers to bother with, then you have let the vandals through the outer perimeter.

The inviolate NLYGD Rule:

Never Let Your Guard Down

Security Awareness

After all the planning work above, there is more. As noted earlier, security must be embedded in the forebrain. One achieves this by implementing an ongoing security awareness program. It takes the form of articles in company and help desk newsletters, posters, stickers on terminals, and newspaper articles about recent virus attacks, computer frauds, and hacker attacks being circulated.

The objective is always the same: to keep security in the forebrain. The message, too, is the same: It *can* happen.

> "So much of security is mind set." GREG SCOTT, *Computing Services Manager, Oregon State University, Corvallis, Oreg.*

Review

- Network security plans fail mainly because people forget they are dealing with a people and not a technology issue.
- An effective network security plan comes to pass because it has management's backing, and because all involved have a say in its formulation. The perception of being arbitrary, high-handed, or insensitive must be avoided at all costs.
- A quick look includes the views of users and network staff.
- Effective security questionnaires are brief, specific, and unbiased.
- A security task force charter gives members direction and a means to achieve their mission.
- There are at least ten areas of scrutiny; they are technical, personnel-related, and administrative in nature.
- A network security plan is half the task.
- Implementation is the other, and harder, half.
- The reason for the difficulty with implementation is that changing peoples' habits takes time.
- High and higher technology provides some obscurity, but the earlier forms are no longer effective.
- *Never, never let your guard down.*

Chapter

14

Where to Start

Objectives

At the end of this chapter you will be able to:

- Itemize specific steps to implement the appropriate level of security in your organization
- Discuss ways to conduct a network audit and why this should be done
- Describe how network discovery software tools work, the types of tools, their advantages and disadvantages, and their limitations
- List four items that constitute the minimum scope of a network audit
- Discuss strengths and weaknesses of SNMP managers as discovery tools
- Explain how data migration happens
- Specify how to secure unattended workstations
- Formulate identification procedures for finding undocumented modems
- Explain how terminal servers shield modems and how to find such modems
- Discuss the role of bridges, routers, and gateways in the discovery process
- Compare automated security audit tools and specify their value to you

- Discuss antipiracy protection to avoid "softlifting"
- Outline the steps in a risk assessment and the desired outcome
- Understand how and why the network security, disaster recovery, and business continuity plans must be viewed as a unified whole
- Indicate the contents and objectives of a physical security review
- Weigh the consequences, if any, of doing nothing

Introduction

The sequence of events in a network security evaluation is shown in Fig. 14.1.

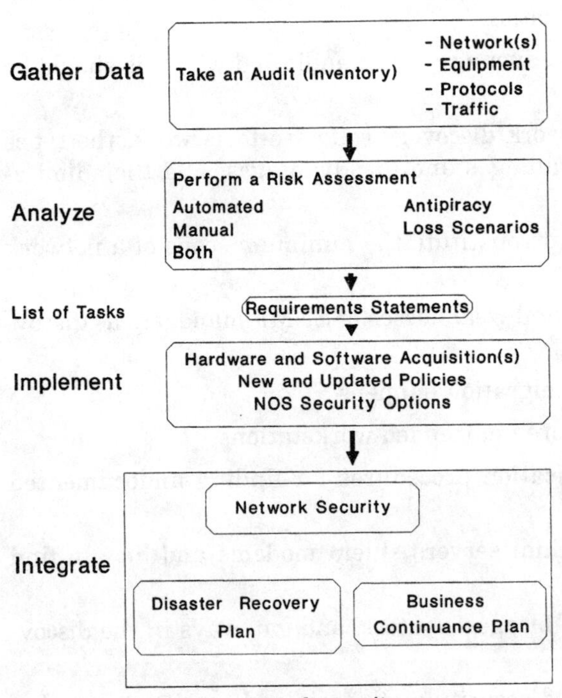

Figure 14.1 Evaluating network security.

Take an Audit

It seems reasonable to say that one cannot build an effective network security plan until one knows the scope of the network. A network audit should minimally include:

- Networks
- Equipment
- Protocols in use
- Traffic flow

Often there are hidden networks of which management may not be aware. There can be unintentional logical links that permit unforeseen communication paths. For instance, a misprogrammed bridge, router, or gateway might permit improper logical links.

Discovery software tools as shown in Fig. 14.2 can explore the network environment. Some just search passively, but others (which are recommended) are more aggressive and interrogate via

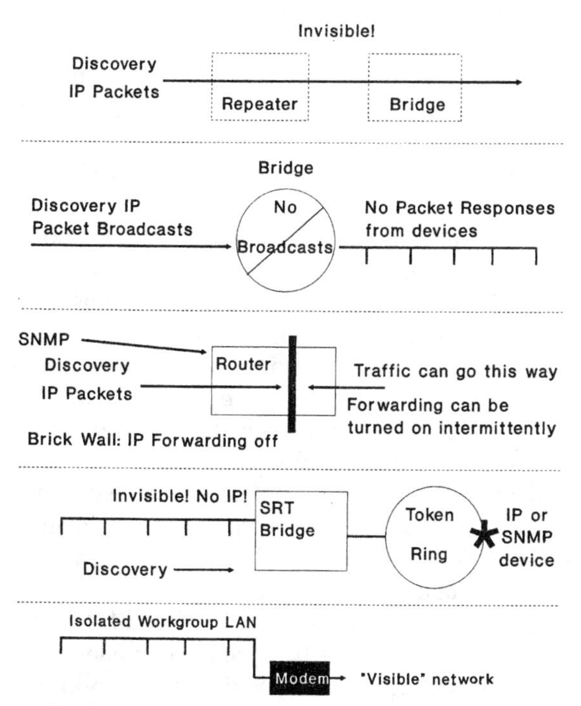

Figure 14.2 Discovery tools and blocks.

broadcast commands. Such commands might include scans for SNMP MIB routing tables or Pings (Ping is a little echo program) to find IP devices.

Products such as Concord Communications' (Marlboro, Mass.) Trakker and NetLabs' (Los Altos, Calif.) Discovery will cross T-1, FDDI, or any link that will carry the protocol under investigation. A weakness of these tools is that routers can block visibility into other network segments unless stations on the remote segments pass data toward the tool on or through the segment to which it is attached.

Bridges and repeaters, working below the network layer, are largely unseen by discovery tools that work mainly at the network (routing) layer. Indeed, bridges that do not support broadcasts may not pass traffic to a segment where a passive discovery tool can see it.

Even active discovery tools can yield incomplete results. If IP forwarding is turned off in a router, IP packets from the discovery tool will not be passed. Yet the router can still pass messages the other way and the router can be turned on or off at will, so a path really does exist.

One partial way around this is to interrogate the router's routing tables under SNMP. Both upstream and downstream SNMP devices can be discerned from the tables.

SNMP-based management systems can help managers find a hidden LAN, but only if at least one device on it is equipped for SNMP. Furthermore, it must be possible to get SNMP packets through a bridge or router in order for the network to be visible to the managing system.

Networks that do not use IP can be a problem, such as token-ring networks that use source routing. A source-route-transparent (SRT) bridge or gateway linking an Ethernet to a token-ring network that lacks SNMP or IP will be invisible, as will the far-side network. The exception is that if any device on the far side is equipped for IP or SNMP, then it will respond to discovery packets.

Novell itself found 1,000 NetWare servers on its internal corporate network. With more than 3,000 employees, that amounts to one server for every three workers. In another case, NetLabs' Discovery found 500 hosts and 10 subnetworks that could gain access into a client's network from Internet locations around the world. The discovery process took ten minutes.

Data can migrate unintentionally as shown in Fig. 14.3 when information is stored on a local database to which other users and databases have access. This violates not only the "need-to-know" principle, but the logical connection could permit the transportation of a virus from machine to machine.

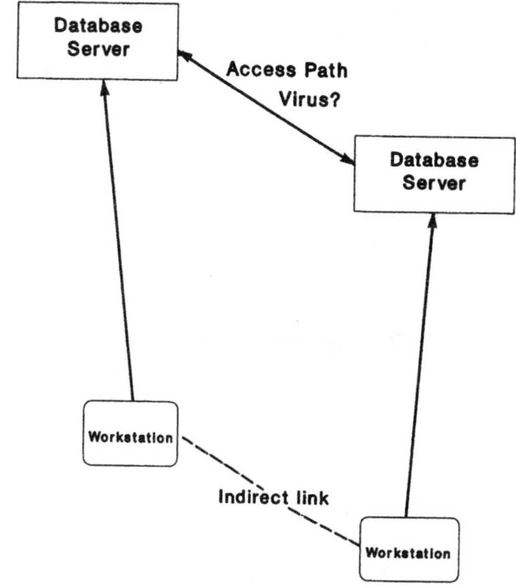

Figure 14.3 Data migration.

Workgroups can purchase their own network equipment and set up a workgroup LAN that is not connected to any other network or that may have an undocumented connection to the primary network via modem.

In sum, a network audit should discover all connected equipment including hardware such as modems and all installed software in each workstation. We discussed this topic more fully in Chapter 7.

A potpourri of protocols

Unanticipated protocols can also show up. At NASA, a discovery tool (Concord Communications' Trakker) unearthed many nonstandard protocols that were later found to be experimental. Unknown protocols, experimental or not, can conflict with authorized protocols and in any case will use some amount of unplanned network bandwidth.

Network traffic

A *traffic study* is necessary because the network should be used only for legitimate purposes. If one person is tying up an internetwork WAN link for personal or frivolous use, that is a breach of network security. Similarly, backups and other bandwidth-intensive activities should be deferred until off-hours when they will not interfere with worker productivity. Traffic studies are essential to the management of network resources. Network managers can then make suitable adjustments.

Improperly set equipment

Sometimes an audit discovers incorrectly set software or active software that should be inactive. For instance, an IBM RS-6000 computer whose routing software is inadvertently left active will pretend to be a router and exchange incorrect routing tables with other routers. Similarly, if a company accidentally leaves its IP routing protocol turned on in an Internet gateway, that makes its network an extension of the Internet.

In Unix and Unix-like systems (more so than DOS) it is possible to move from a less-secure application to one with greater security by using the less-secure machine to help obtain a higher level of access. Typically, this kind of security one-upmanship requires an insider or ex-insider who knows what he or she is doing. Similarly, certain applications have privileged programs. If adversaries can gain control of the system, they can modify the privileged program and bypass security defenses.

A characteristic supposition in both the above cases is that the insider or ex-insider has or retains password access. Former employees sometimes become curious to see if they can still get in, and that may beget malicious thoughts. Eliminate this potential risk by invalidating passwords for electronic mail, voice mail, and all other systems the moment the employee departs, or just beforehand.

Some companies do not protect electronic mail by password because they consider it so innocuous. Yet if internal information or confidential data belonging to customers reside on the electronic mail system, then owners have a legal fiduciary responsibility to protect it.

Unattended workstations

A logged-in but unattended workstation is an invitation to hacking. Pyramid Software's (Rocky Hill, Conn.) PC/DACS automatically logs out any PC that has been inactive for a selected time. Then the user must log in again. A similar product, NetOFF (Citadel Software, Phoenix, Ariz.), logs users out after a period of time from 1 second to 24 hours. It includes a screen blanker with password reentry.

One enterprising programmer wrote a program to get around this perceived annoyance. He wrote a small TSR program that generated a carriage return every 29 minutes. The lesson here is that people—programmers, in particular—must be taught to respect the need for security. As we have said consistently, security is a people issue.

As a matter of policy, you may wish to instruct users to log off databases or applications before they leave the area. Security-sensitive users will go so far as to redirect their operating systems away from network links or shared remote devices before leaving their workstations unattended.

If users are made aware that they are held accountable for system usage under their password, then they will have a vested interest in maintaining security. If no one is watching, why should they care?

Undocumented modems

Small, inexpensive, and often buried in a PC slot, modems are hard to spot, especially since their use is transitory. When not in use, they are invisible except to an audit tool such as PC Census. They are so inexpensive that sometimes users will buy them with petty cash or their own money. Thus they are even invisible to a purchase order review.

Access can be indirect. If your computers are connected to a packet-switching network, the Internet or IBM's Infonet, for example, then a hacker has a path to your front door. Keep the door closed by changing passwords often and opening the door just enough to admit a legitimate user while rejecting all others, as with time-of-day router enabling.

To prevent users from installing their own modems, software is available to lock up PC communications ports and serial interfaces while not restricting the machine's local processing ability. One such product is Cordant's (formerly Centel Federal Systems) Assure PC security system.

Beside auditing, modem calling patterns can show up in PBX call management system reports. Dial-up calls are often lengthy, so an unusually long call can be a clue. Sometimes these sessions will occur at odd times, especially if the employee is civil enough to know that calling before 8 A.M. or after 5 P.M. is cheaper.

PBXs can be programmed to prevent calls, disconnect calls in progress, or restrict the direction and duration of a call during off-hours. Both known and unknown modems can be managed indirectly by limiting their access to the switched network.

Security can be added to modems by purchasing security modems that include dial-back capability, password access control, or even DES encryption.

Discovery tools cannot find modems because they have neither IP nor SNMP addresses. Furthermore, modems can be shielded by terminal servers that reside between the discovery tool and the modem, as shown in Fig. 14.4. The key is to look at the terminal server's ports for lengthy activity or repeated failed login attempts.

Another method is to go back to the source. Using this method, a monitor collects the IP addresses of anyone attempting to access the file server's data, whether successful at login or not.

Another Cordant product, server-based, authenticates the caller into the server. If the PC does not supply the correct authenticator, the PC is locked up. That will stop an unauthorized user from accessing a server. It also forces a legitimate user to register their modem if they seek access to the server or network.

Bridges, routers, gateways, and switches

If you are able, set these devices to cause an alarm upon registering an unauthorized address or unusual activity. Set internal time controls to open and close access, especially for Internet links. Information flow can be bidirectional or unidirectional as desired. It is also possible to go out into the larger network yet prevent

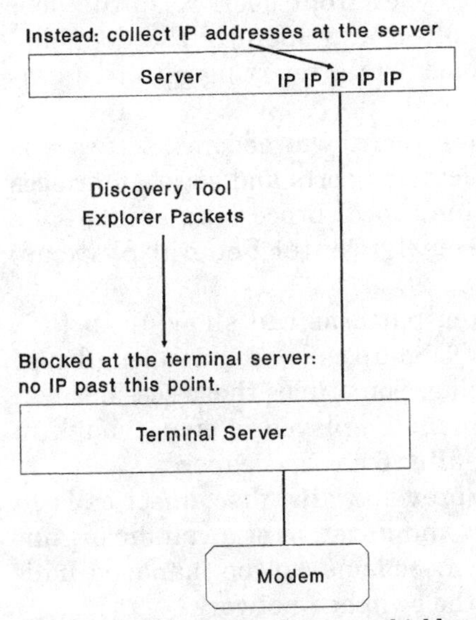

Figure 14.4 How terminal servers shield modems.

access the other way. Learning bridges, for example, can invalidate old addresses and request human approval to reestablish them.

Serial Line Internet Protocol

The Serial Line Internet Protocol (SLIP) is commonly used in TCP/IP environments, especially for dial-in to Ethernet LANs from remote workstations to hosts (servers). SLIP is widely available and easily implemented. Basically, all it does is send an IP datagram, byte by byte, typically at speeds from 1,200 to 19,200 bits/s. Although SLIP has no speed limit, these are popular dial-in speeds. Both start–stop and synchronous protocols are supported, but SLIP itself contains no error checking.

If a gateway attached to the LAN has enabled IP forwarding, then the dialer can call the gateway and gain access to any internal LAN, workstation, or Internet-connected LAN. The presence of this protocol in the network should be a flag, as SLIP is a common vehicle for hacking.

Audit Methodology

Do an audit periodically, at least once a year. Start from the network's information center and work outward. Analyze paths (Fig. 14.5) by which information can be accessed, both obvious and nonobvious, such as by stealing discarded printouts.

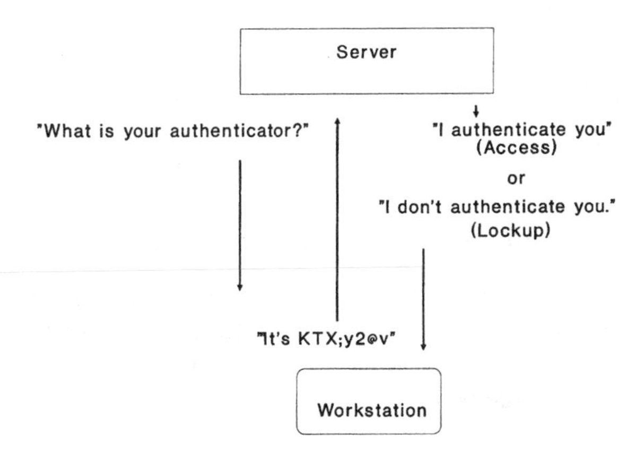

Dial-in to Ethernet often uses the SLIP protocol.

Figure 14.5 Server authentication.

Logically enough, decentralized organizations are more likely to have hidden networks and equipment than strongly centralized ones. Decentralized organizations have decentralized security, which is to say inconsistent security. Security should be addressed at the global, enterprise level, not at the local level, as we have said before. In a decentralized system, the overall security will be no better than the weakest security subsystem implementation ("a chain is only so strong as its weakest link").

Be sure to review purchase orders for computers, equipment, and software. Even hidden systems have to be bought, and their purchase orders may give you a paper audit trail back to many hidden components and even isolated networks.

Automated security audits

Software tools exist to lend structure to a security audit. Typically, such tools:

- Ask probing security-related questions
- Assign security-related priorities
- Recommend corrective actions and track the progress of these actions toward resolution

The scope of a good tool will include at least:

- Physical access controls
- Data communications
- Personnel security
- Data security
- Audit and application controls
- Disaster recovery
- Business continuation planning tools

One such tool is *Information Systems Security Audit,* offered by Contingency Strategies Associates, Inc. of Avon, Conn. At $300 a copy or $1,000 for a site license, it is a very good start and will stimulate your own security ideas as well.

Antipiracy protection

Your organization terminates an employee. Disgruntled, they dial the Software Publishers' Association's (SPA) 800-number Piracy

Hot Line. The SPA then files a copyright infringement against your organization. The following day, armed with a court order or an injunction, SPA shows up at your door with an armed U.S. Marshal for company. It can happen. SPA has sued or raided over 100 companies in the past three years. An average of two violators a week are prosecuted.

It happened to catalog retailer DAMARK International, Inc. of Minneapolis, Minn. DAMARK reached an out-of-court settlement with SPA after a distracting and meddlesome investigation. DAMARK paid legal and software license fees. They lost valuable work time in the process and were forced to make a contribution to the SPA's Copyright Protection Fund.

All it takes is one phone call to get the "software police" out after you. To avoid being accused of "softlifting" implement the following policies (see Table 14.1):

- Set up a policy prohibiting the duplication of copyrighted software or the use of unlicensed copies of such software.
- Insist on having a set of original user manuals at each workstation where that software is installed. There should be a set of original disks (stored away) and manuals for each set of hard-disk installed software.
- Centralize the purchase, installation, and registration of all PC software in one place in your organization.

TABLE 14.1 Antipiracy Checklist

- ✓ Set and enforce a no-duplication policy.
- ✓ Require original software manuals at each workstation where the software is installed.
- ✓ Centralize software purchase, distribution, and registration.
- ✓ Educate managers and users.
- ✓ Perform an initial audit.
- ✓ Document purchases to date.
- ✓ Register all purchases with the vendor.
- ✓ Destroy software for which you cannot prove ownership.
- ✓ Make backups of purchased software, but store both them and the originals in a place where they cannot be duplicated casually.
- ✓ Mark all backups as containing copyrighted material.
- ✓ Repeat the audit annually.

As usual, a little planning can avoid a catastrophe:

- Educate management. This should take the form of a memorandum with the proposed policy after review by counsel.
- Make the acquisition of software the responsibility of one person or organizational unit. This pinpoints accountability.
- Audit your software. SPAudit is an SPA software tool that finds about 700 SPA-member products. It gathers and prints an index of executable programs that you actually have.

For a copy of this index, you can write the Software Publishers Association, 1730 M Street NW, Suite 700, Washington, D.C. 20036-4510. You can also call (202) 452-1600 and ask for a copy of the Self-Audit Kit.

You may wish to think carefully before calling. Once you call, you have identified yourself to SPA as having a possible problem. It seems inadvisable to make yourself a voluntary target. Instead, get a copy through a neutral third party, your attorney, or use another tool.

Further steps to take are the following:

- Assemble documentation that proves purchase. If necessary, go back to your vendor for proof of purchase.
- Be sure your software is registered with the vendor. Vendors will supply you with a list of your registered software.
- Think about standardizing on one set of software for each function, such as spreadsheets and word processing.
- Destroy any software for which you cannot show proof of purchase, including hard and floppy disk copies. Replace with purchased software and register promptly. Try to use a centralized, systematic installation process. Be sure that employees know about the deletion before it is removed. You could be improperly deleting software they purchased personally.
- Make backup copies and label "Not for Duplication—contains copyrighted material." Store originals and duplicates in a secure area.
- Audit software as well as hardware annually.

Risk Assessment

Once an inventory is complete, you then have enough data to perform a *risk assessment*. A risk assessment produces a set of security requirements unique to the organization. Clearly, every organization has its own business needs, its own installed base of computing and communications equipment, its own types and trustworthiness of its users, and the like.

The assessment makes use of past history, what insurance underwriters call *loss experience*. Losses indicate weaknesses that need fixing.

Perform the inventory and risk assessment at the global, enterprise level. Otherwise your security measures will be weaker in one area than another, creating opportunities for mischief.

By the same token, apply security measures to all applications. The global approach simplifies the user interface, adds consistency, and helps make security more transparent to the user (example: encrypt everything, not specific applications).

Potential loss scenarios

Ask yourself some uncomfortable questions, as posed in Table 14.2. You may recall that estimating the frequency and multiplying by the dollar loss can help prioritize loss scenarios.

Requirement statements

For each loss scenario, a requirement statement indicates what is needed to prevent the loss or at least make it recoverable. For instance, if physical access to a server makes its theft, tampering, or other interference possible, placing it under lock and key is the appropriate requirement statement. Each scenario may have several or even many requirements to prevent or neutralize the scenario. The sum of requirement statements represents the useful output of the risk assessment process.

The easy part

Requirement statements indicate what, if any, hardware, software, or other security measures need to be taken. *Understand what you*

TABLE 14.2 Potential Loss Scenarios—Some Uncomfortable Questions

Who or what could cause a loss?
What magnitude could it take?
Why might the loss occur?
What systems might be involved?
What could the possible consequences be?

need before you spend money. Thus you cannot be diverted or closed-minded when evaluating products. You will *know* what you are looking for.

The outcome of the risk assessment should also determine how many of the NOS security features need to be enabled. It is inappropriate to turn them all on if some portions are not needed.

Use internal security features if possible. According to a study by SRI International and commissioned by IBM, internal features cost a tenth of what it costs to bolt additional systems onto an existing one.

Employ add-on systems only if necessary and avoid building your own software, unless there is absolutely no other option. If you can find 80 percent of what you need in a purchased product, buy it.

Bolt-on security packages

If you decide that you need additional security software, look for the features itemized in Table 14.3.

Business continuity planning

A disaster recovery plan helps an organization cope with a catastrophe. It does not provide a blueprint for the timely resumption of business. That is what a continuity plan does.

Continuity planning reduces confusion, improves communication between the business elements affected, and reduces the time needed to get the business going again. Thus the focus is chronologically beyond that of a disaster recovery plan, but no less essential.

Vendors of disaster recovery services such as Comdisco (Foster City, Calif.) and SunGard (Philadelphia, Penn.) offer continuity planning services as a logical adjunct to their well-known recovery

TABLE 14.3 Bolt-On Security Package Feature Checklist

✓ User activity audit trails and customizable report formats therefrom

✓ Password-restricted boot control

✓ The ability to restrict executable commands via menus or otherwise

✓ DES encryption of cable traffic (optional)

✓ DES encryption of hard disk and floppy disk files

✓ Directory- and file-level access control via passwords

✓ Screen blanking and keyboard locks after a period of inactivity or if initiated by the user

✓ Virus scanning via CRC checksums or signature scanners

services. For example, Comdisco sells a software tool called ComPAS (for Comdisco Plan Automation System). Running on an IBM-PC or compatible platform, a complete continuity plan can be built from it.

Since the tool is a Comdisco product, expect it to recommend Comdisco-offered services. It *is,* however, a useful tool when employed with one's eyes open. Similarly, SunGard Planning Solutions (of Wayne, Penn.) offers consulting services and seminars to help businesses develop a business recovery plan.

The Cost of Doing Nothing

There might be no cost, and in fact, a savings. You might beat the odds and never be struck by a virus, a hacker, a disaster, or any of the other unpleasantry presented in this book for the rest of your career.

It is a virtual certainty that some event will occur that necessitates an emergency response. It is part of your managerial responsibility to be prepared for the day that happens.

We have tried to present practical information and approaches that you can implement starting today. *Do it all.*

Security Checklists

To help you get started, we have summarized the recommended actions given throughout this book in the remainder of this chapter, concentrating on the following:

- Physical security in Tables 14.4 and 14.5
- Access security in Table 14.6
- Media security in Table 14.7
- Personnel and security policy in Table 14.8

Physical security

A complete physical security review consists of far more than just network security. It includes the items in Table 14.4. Table 14.5 is a checklist that adds further detail.

Review

- Start by performing a network inventory. You must know what you have before you can go further.

TABLE 14.4 Physical Security

✓ *An access control method review,* both physical (doors, access locks, etc.) and logical (user IDs, passwords, password generation, and auditing and password levels).
✓ *Fire detection and protection systems* such as sprinklers and Halon (Halon is being replaced by other more environmentally acceptable chemicals).
✓ *Power protection,* including filtering and uninterruptible power supplies for essential equipment.
✓ *Construction specifications,* specifically earthquake protection, heating, ventilation, and air conditioning (HVAC) requirements, and floor load limits.
✓ *External physical evaluation* such as flood exposure, proximity to explosive sources (such as fuel tank farms), proximity to airport landing and departure paths, and exposure to severe weather such as hurricanes and tornadoes.

- Consider using a security audit tool. It can help in the next step, a risk assessment.
- The outcome of a risk assessment is a set of requirement statements designed to improve security in areas where it has been found deficient.
- Requirement statements lead to policy changes, software and/or hardware acquisitions, and appropriate security settings in NOS software.
- Bolt-on security packages are expensive. Use what you have if possible before considering add-on services.
- Avoid writing your own code if you can get 80 percent or more of what you need in commercial software.
- Take the global, enterprise-wide view toward security.
- Do not forget that illegal copying of software places your network at just as much risk as if it were assaulted by a hacker. In effect, you become a hacker.
- Business continuity planning is the logical and chronological extension of the disaster recovery plan. It specifies what activities must continue and how they will be done in a recovery posture.
- It is possible, but foolish, to do nothing. It is a certainty that you can improve network security without spending a dime by just using what you have and implementing policies that minimize your exposure.
- Checklists, while not the ultimate security tool, do give you some direction if you must do something *now.*
- Four checklist areas are covered: physical security, access controls, media security, and personnel and policies.

TABLE 14.5 Physical Security Checklist

- ✓ Servers physically isolated.
- ✓ Server console locked.
- ✓ A: drive disabled.
- ✓ Hubs enclosed and locked.
- ✓ Wiring secured.
- ✓ LAN analyzers locked.
- ✓ LAN analyzers' password protected.
- ✓ Workstations not open to the public.
- ✓ Workstation automatic screen blanking and keyboard locking.
- ✓ Workstation automatic logout.
- ✓ Disable A: drives in workstations.
- ✓ Turn monitors so they cannot be seen from the outside.
- ✓ Use a fax key operator to send and receive faxes.
- ✓ Designate a secure fax machine that only the key operator can use.
- ✓ Better yet, get a fax server and eliminate the problem altogether.
- ✓ Add an uninterruptible power supply to all servers and critical routers, bridges, and hubs.
- ✓ Filter all power to hubs and workstations.
- ✓ Provide TEMPEST shielding where justified.
- ✓ Require employees to identify themselves via a badge system.
- ✓ Visitors should receive red badges and should be escorted.
- ✓ Provide employees handling sensitive information with office safes.
- ✓ Check all fire extinguishers at least annually.
- ✓ Install sprinkler systems, Halon, or other fire suppressant systems.
- ✓ If a concern, evaluate the building for its ability to withstand severe weather, flooding, or earthquake.
- ✓ Adopt and implement a backup rotation plan that keeps recent, essential data off-site.
- ✓ If justified, consider a real-time off-site backup system as the brokerages use.
- ✓ Consider the impact of a long-term electrical or communications failure.
- ✓ Investigate and retain a disaster recovery service if necessary.
- ✓ Consider the network security plan, the disaster recovery plan, and the business continuance plan as unified elements of a survival plan.
- ✓ Implement physical security for everyone, not just for small groups.
- ✓ Review your business insurance policy to see what is covered and what is not. Review with the agent the documentation or justification needed to process a claim successfully.
- ✓ Review with counsel any fiduciary or contractual responsibilities that could jeopardize the company if it cannot function; look at both short-term and long-term liabilities.

TABLE 14.5 Physical Security Checklist (*Continued*)

- ✓ For physical security, take the perimeter approach. Start from the outside and work inward. The objective is to prevent intrusion.
- ✓ Multiple barriers provide redundancy, weaken an adversary, and have a deterrent quality.
- ✓ Multiple barriers may be physical, as with an outer fence and inner fence. They may also be logical, as with multiple passwords.
- ✓ Be low-key. Make the building look like an abandoned warehouse. Done properly, no one will bother to look twice.
- ✓ Buildings should be windowless. Windows are real weak spots.
- ✓ Walls should extend right to the roof, as do firewalls. A gap leaves a space for a penetrator.
- ✓ Locks should be of the "twisting tumbler" or circular Ace-type where deemed necessary. Blanks are hard to find and picking is nearly impossible with these types of locks.
- ✓ If electrical locks are used, shield pushbuttons so codes cannot be seen.
- ✓ Except for trained professionals, garden-variety guards are as much of a risk if not more than a perpetrator.
- ✓ If guards are used, use a time clock and key system.
- ✓ Patrol frequency and effectiveness are closely tied together. Keeping guards moving gives them less time to think up mischief.
- ✓ Irregular patrol frequencies are very desirable. Routine is bad.
- ✓ Remember that human beings do not do well at repetitive tasks.
- ✓ Light all areas well, leaving as little shadow as possible. Lights are also an excellent deterrent.
- ✓ For extreme-security environments, consider use of automated entry booths, so-called "man traps."
- ✓ If guards are used, consider providing encrypted communications.
- ✓ Guards and all the above add psychological deterrence.
- ✓ Designate specific individuals who can bring equipment in and out. Passes are useless.
- ✓ Employ alarm systems and closed-circuit TV as required.
- ✓ Information in a wastebasket is not "destroyed." Use a shredder.
- ✓ Promote questioning by staff of people who look out of place.

TABLE 14.6 Access Security Checklist

- ✓ Implement physical access controls as above.
- ✓ Access is a matter of authentication. Consider using two-part authentication (e.g., something one *has* plus something one *knows*).
- ✓ Implement user IDs and passwords for all systems, including electronic mail.
- ✓ Permit users to pick from an assortment of supplied phonetic passwords.
- ✓ Activate weak and stale password protection features in NOSs.
- ✓ Use at least six-character passwords, mixing numbers and letters.
- ✓ Consider the use of a dynamic password that changes every minute or so.
- ✓ Use multiple passwords if necessary to gain access to the most sensitive data.
- ✓ Implement NCP Packet Signature authentication in any version of NetWare.
- ✓ Do not permit unencrypted passwords (e.g., NetWare can do so if instructed).
- ✓ Use intruder detection and time-of-day access controls in NOSs.
- ✓ Install an audit system that tracks logins and logouts.
- ✓ Avoid logging in as supervisor whenever possible.
- ✓ Set rights and attributes so that they are neither too generous nor too restrictive. If in doubt, make them tighter and loosen them as experience dictates.
- ✓ Use an audit tool to detect weak right and attribute combinations.
- ✓ Monitor all external connections via modem, the Internet, and private networks such as Infonet or BT Tymnet.
- ✓ Remove all unnecessary protocols from routers. Set time-of-day limits on routing. Permit outbound but not inbound connections as needed.
- ✓ Dial-back has its limits as an authenticator. Be aware of them (e.g., taking over a session once established).
- ✓ Remove obvious ("LUBE") and default passwords from all systems.
- ✓ Consider encryption or authentication on exposed data circuits or wireless links of any type.
- ✓ Establish stiff penalties for writing down or sharing passwords.
- ✓ Have employees sign security statements annually.
- ✓ Preestablish login paths to discourage browsing.
- ✓ Consider using secure or switching LAN hubs.
- ✓ Use a discovery or audit tool to find modems and the real network.
- ✓ Try hacking your own system. Get creative.

TABLE 14.7 Media Security Checklist

- ✓ Forbid the use of shareware or downloading executable files from bulletin boards to a workstation connected to a network until such files have been scanned for viruses. Use two or more scanners to be really sure.
- ✓ Scan on a standalone machine.
- ✓ Prohibit bringing in disks from elsewhere until they have been scanned.
- ✓ Have vendors demonstrate their wares on their own equipment.
- ✓ Scan all disks, even those that go in routers and bridges, if they are in DOS format.
- ✓ Use a virus scanner on the server and one in each workstation.
- ✓ Remember that Norton Utilities' Disk Doctor and others can read disks directly. Limit distribution of such utilities, and their presence on a workstation during an audit should be a warning flag if the person does not seem to need it.
- ✓ Use Norton Wipedisk or equal to really randomize "erased" disks.
- ✓ Use a similar tool to randomize hard disk erasures.
- ✓ Do not permit the movement of a diskette from a less-secure network to a more-secure network. It could bypass all the secure network's security features.
- ✓ Consider using a signature system to verify that a disk file has not been tampered with.
- ✓ Consider using diskless PCs or dumb terminals.
- ✓ Consider remote booting via ROM and server boot file.
- ✓ Establish a systematic media backup rotation plan, including off-site storage.
- ✓ Find and retain the names of firms who do media restoration, as from damaged hard disks or wet-soaked diskettes. Keep their phone numbers handy.
- ✓ Find out if you have media insurance. Such insurance usually covers the cost of reconstructing software and information.
- ✓ Write-protect all possible diskettes to prevent virus transfers.
- ✓ Diskettes can be damaged by temperatures of 125°F or so, which can easily be reached in a closed car on a summer day or near an oven.
- ✓ Magnetic fields can damage disks. A field strength of 50 oersteds is sufficient, about equal to that of a paper clip dispenser.
- ✓ Paper clips and floppy disks should never be used together.
- ✓ Except for felt-tipped pens, writing on 5¼ in floppy disks can damage them.
- ✓ Diskettes can be copied, bent, stolen, or soiled. They also burn easily.

TABLE 14.8 Personnel and Policy Checklist

- ✓ People are an organization's greatest asset. Their collective formal training and on-the-job experience is an investment beyond price.
- ✓ Replacing a trained employee can be costly and disruptive.
- ✓ Treating people properly raises their productivity and positively disposes them toward the security program, which is seen to include them.
- ✓ Treating people properly reduces the likelihood of litigation or theft by an employee. According to a DPMA study, about 20 percent of computer crimes are acts of revenge.
- ✓ It should also be recalled that the threat from an insider far exceeds that from an outsider.
- ✓ Temporary personnel are a greater potential threat than permanent staff.
- ✓ Conduct a background investigation of prospective employees. A fifth of computer crimes are committed by those having a criminal record.
- ✓ If legal, conduct a polygraph test. Conduct psychological testing if the position's authority level justifies it.
- ✓ Employees in critical positions, such as network and database administrators, should be watched for signs of discontent or duress. They should be placed on paid administrative leave and their privileges canceled until the matter is resolved.
- ✓ Terminated employees should have their privileges revoked immediately.
- ✓ Divide duties and responsibilities so there are checks and balances. Collusion should be required to cause significant damage, although collusion is a factor in about half such cases and is no magic bullet.
- ✓ Employ psychological deterrence wherever possible and in any legitimate form.
- ✓ Monitoring techniques, including audits, periodic reinvestigation, and simple supervisory awareness are all required. Most computer crimes are discovered by accident because someone noticed something out of order.
- ✓ Administrative policies define for people what is expected of them.
- ✓ Implement the "need-to-know" principle across the board.
- ✓ Insist that employees take their vacations when due.
- ✓ Rotate job positions periodically.
- ✓ Use signed security agreements, either general or specific in nature.
- ✓ Spell out security policies in clear, concise terms.
- ✓ Update the policies as needed; annually is usually sufficient.
- ✓ Conduct annual training to sensitize employees.
- ✓ Upper management must endorse security policy and set the example in its implementation.
- ✓ Conduct an ongoing security awareness program.
- ✓ Urge employees to report back doors they find or suspicious calls that sound like hackers using "social engineering" to gain access.
- ✓ Use new-hire security indoctrination.

TABLE 14.8 Personnel and Policy Checklist (*Continued*)

✓ Set up tailored sessions for specific audiences.
✓ Get attention with bulletins, posters, and so on.
✓ Publicize recent security incidents, whether internal or external.
✓ Amend existing policies rather than create new ones.
✓ Work with those who will have to live with the security policy or it will be ignored.

- Do not wait. Start today. Security problems in our society will get worse before they get better.

Appendix. Vendors

Inventory tools find network extensions and inventory network-attached devices. Below is a partial list of vendors:

Vendor	Product
Blue Lance	LT Auditor for NetWare ($700)
Brightwork Development, Inc.	LAN Automatic Inventory for NetWare and LAN Manager
Certus International, Inc. Cleveland, Ohio	Certus LAN
Concord Communications, Inc. Marlboro, Mass.	Trakker (discovery) network monitoring system
DEMAX Software	SecureMax for DEC VAX and SunOS Unix network security DeMax Distributed Security Technology for VAX, Unix, or NetWare System Detective for VAX VMS systems; spots intruders; reports them; captures their sessions; locks keyboards and freezes screens
Frye Computer Systems, Inc. Boston, Mass.	LAN Directory (Novell, PC, MAC networks; $500)
Horizons Technology	LAN Auditor for DOS, Windows, NetWare, VINES; LAN Manager
The LAN Support Group Houston, Tex.	Bindview + for NetWare; includes Node Information Manager Inventory Monitor

MAGEE Enterprises, Inc. Norcross, Ga.	Network HQ; $4,500; tracks NetBIOS, Macintosh, NetWare, LAN Manager, VINES, NDIS, ODI configurations
Multima Corp. East Greenwich, R.I.	NetKeeper for DOS and NetWare networks
NetLabs, Inc. Los Altos, Calif.	Discovery network management tool
Preferred Systems, Inc. West Haven, Conn.	PSI: Origen for NetWare server management
Tally Systems Corp. Hanover, N.H.	PC Census
TouchStone Software Corp. Huntington Beach, Calif.	CheckIt LAN
Westinghouse Communications Software	Communications Management Systems

Glossary

access A specific type of interaction between a subject and an object that results in the flow of information from one to another.

access control The process of limiting access to the resources of a system only to authorized programs, processes, or other systems (in a network). Synonymous with **controlled access** and **limited access.**

access control mechanism Hardware or software features, operating procedures, management procedures, and various combinations of these designed to detect and prevent unauthorized access and to permit authorized access in an automated system.

access level The hierarchical portion of the security level used to identify the sensitivity of data and the clearance or authorization of users. *Note:* The access level, in conjunction with the nonhierarchical categories, forms the *sensitivity label* of an object. See **category, security label,** and **security level.**

access list A list of users, programs, and/or processes and the specifications of access categories to which each is assigned.

access period A segment of time, generally expressed on a daily or weekly basis, during which access rights prevail.

access port A logical or physical identifier that a computer uses to distinguish different terminal input–output data streams.

access type The nature of an access right to a particular device, program, or file (e.g., read, write, execute, append, modify, delete, or create).

accountability The property that enables activities on a system to be traced to individuals who may then be held responsible for their actions.

Excerpted and adapted from NCSC-TG-004.

add-on security The retrofitting of protection mechanisms, implemented by hardware or software.

administrative security The management constraints and supplemental controls established to provide an acceptable level of protection for data. Synonymous with **procedural security.**

advanced authentication The use of automated one-time passwords between machines or between a user and a machine such as a security card (token) in order to authenticate a user or the identity of another machine. Advanced authentication excludes the use of conventional passwords.

assurance A measure of confidence that the security features and architecture of an automated information system accurately mediate and enforce the security policy.

attack The act of trying to bypass security controls on a system. An attack may be active, resulting in the alteration of data, or passive, resulting in the release of data. *Note:* The fact that an attack is made does not necessarily mean that it will succeed. The degree of success depends on the vulnerability of the system or activity and the effectiveness of existing countermeasures.

audit trail A chronological record of system activities that is sufficient to enable the reconstruction, reviewing, and examination of the sequence of environments and activities surrounding or leading to an operation, a procedure, or an event in a transaction from its inception to final results.

authenticate (1) To verify the identity of a user, device, or other entity in a computer system, often as a prerequisite to allowing access to resources in a system. (2) To verify the integrity of data that have been stored, transmitted, or otherwise exposed to possible unauthorized modification.

authenticator The means used to confirm the identity or verify the eligibility of a station, originator, or individual.

authorization The granting of access rights to a user, program, or process.

automated security monitoring The use of automated procedures to ensure that security controls are not circumvented.

availability of data The state when data are in the place needed by the user, at the time the user needs them, and in the form needed by the user.

back door Synonymous with **trap door.**

backup plan Synonymous with **contingency plan.**

between-the-lines entry Unauthorized access obtained by tapping the temporarily inactive terminal of a legitimate user. See also **piggyback.**

browsing The act of searching through storage to locate or acquire information without necessarily knowing of the existence or the format of the information being sought.

callback A procedure for identifying a remote terminal. In a callback, the host system disconnects the caller and then dials the authorized telephone number of the remote terminal to reestablish the connection. Synonymous with **dial-back.**

capability A protected identifier that both identifies the object and specifies the access rights to be allowed to the accessor who possesses the capability. In a capability-based system, access to protected objects such as files is granted if the would-be accessor possesses a capability for the object.

category A restrictive label that has been applied to classified or unclassified data as a means of increasing the protection of the data and further restricting access to the data.

communications security (COMSEC) Measures taken to deny unauthorized persons information derived from telecommunications of the U.S. government concerning national security and to ensure the authenticity of such telecommunications. Communications security includes cryptosecurity, transmission security, emission security, and physical security of communications security material and information.

compromise A violation of the security policy of a system such that unauthorized disclosure of sensitive information may have occurred.

compromising emanations Unintentional data-related or intelligence-bearing signals that, if intercepted and analyzed, disclose the information transmission received, handled, or otherwise processed by any information processing equipment. See **TEMPEST.**

computer abuse The misuse, alteration, disruption, or destruction of data-processing resources. The key aspect is that it is intentional and improper.

computer cryptography The use of a crypto-algorithm in a computer, microprocessor, or microcomputer to perform encryption or decryption in order to protect information or to authenticate users, sources, or information.

computer fraud Computer-related crimes involving deliberate misrepresentation, alteration, or disclosure of data in order to obtain something of value (usually for monetary gain). A computer system must have been involved in the perpetration or coverup of the act or series of acts. A computer system must have been involved through improper manipulation of input data; output or results; applications programs; data files; computer operations; communications; or computer hardware, systems software, or firmware.

computer security subsystem A device designed to provide limited computer security features in a larger system environment.

concealment system A method of achieving confidentiality in which sensitive information is hidden by embedding it in irrelevant data.

confidentiality The concept of holding sensitive data in confidence, limited to an appropriate set of individuals or organizations.

configuration control The process of controlling modifications to the system's hardware, firmware, software, and documentation that provides sufficient assurance that the system is protected against the introduction of improper modifications prior to, during, and after implementation. Compare with **configuration management.**

configuration management The management of security features and assurances through control of changes made to a system's hardware, software, firmware, documentation, test, test fixtures, and test documentation throughout the development and operational life of the system. Compare with **configuration control.**

confinement The prevention of the leaking of sensitive data from a program.

contamination The intermixing of data at different sensitivity and need-to-know levels. The lower-level data are said to be contaminated by the higher-level data; thus, the contaminating (higher-level) data may not receive the required level of protection.

contingency plan A plan for emergency response, backup operations, and post-disaster recovery maintained by an activity as a part of its security program that will ensure the availability of critical resources and facilitate the continuity of operations in an emergency situation. Synonymous with **disaster plan** and **emergency plan.**

control zone The space, expressed in feet of radius, surrounding equipment processing sensitive information that is under sufficient physical and technical control to preclude an unauthorized entry or compromise.

controlled access See **access control.**

controlled sharing The condition that exists when access control is applied to all users and components of a system.

cost–risk analysis The assessment of the costs of providing data protection for a system versus the cost of losing or compromising the data.

countermeasure Any action, device, procedure, technique, or other measure that reduces the vulnerability of a threat to a system.

crypto-algorithm A well-defined procedure or sequence of rules or steps used to produce a key stream or cipher text from plain text and vice versa.

cryptography The principles, means, and methods for rendering information unintelligible and for restoring encrypted information to intelligible form.

cryptosecurity The security or protection resulting from the proper use of technically sound cryptosystems.

Data Encryption Standard (DES) A cryptographic algorithm for the protection of unclassified data, published in Federal Information Processing Standard (FIPS) 46. DES, which was approved by the National Institute of Standards and Technology, is intended for public and government use.

data integrity The property that data meet an a priori expectation of quality.

data security The protection of data from unauthorized (accidental or intentional) modification, destruction, or disclosure.

degaussing The erasure of magnetic media by the application of an electromagnetic field sufficient in strength to destroy any data residing on the medium.

denial of service Any action or series of actions that prevent any part of a system from functioning in accordance with its intended purpose. This includes any action that causes unauthorized destruction, modification, or delay of service. Synonymous with *interdiction*.

dial-back Synonymous with **callback.**

dial-up The service whereby a computer terminal can use the telephone to initiate and effect communication with a computer.

disaster plan Synonymous with **contingency plan.**

discretionary access control A means of restricting access to objects based on the identity and need to know of the user, process, and/or groups to which they belong. The controls are discretionary in the sense that a subject with a certain access permission is capable of passing that permission (perhaps indirectly) on to any other subject. Compare with **mandatory access control.**

domain The unique context (e.g., access control parameters) in which a program is operating; in effect, the set of objects that a subject has the ability to access. See also **process** and **subject.**

embedded system A system that performs or controls a function, either in whole or in part, as an integral element of a larger system or subsystem.

emergency plan Synonymous with **contingency plan.**

emission security The protection resulting from all measures taken to deny unauthorized persons information of value that might be derived from intercept and from an analysis of compromising emanations from systems.

end-to-end encryption The protection of information passed in a telecommunications system by cryptographic means, from point of origin to point of destination.

entrapment The deliberate planting of apparent flaws in a system for the purpose of detecting attempted penetrations.

environment The aggregate of external procedures, conditions, and objects that affects the development, operation, and maintenance of a system.

erasure A process by which a signal recorded on a magnetic media is removed. Erasure is accomplished in two ways: (1) by alternating current erasure, by which the information is destroyed by applying an alternating high and low magnetic field to the media; or (2) by direct current erasure, by which the media are saturated by applying a unidirectional magnetic field.

fail-safe Pertaining to the automatic protection of programs and/or processing systems to maintain safety when a hardware or software failure is detected in a system.

fail-soft Pertaining to the selective termination of affected nonessential processing when a hardware or software failure is detected in a system.

failure access An unauthorized and usually inadvertent access to data resulting from a hardware or software failure in the system.

failure control The methodology used to detect and provide fail-safe or fail-soft recovery from hardware and software failures in a system.

fault A condition that causes a device or system component to fail to perform in a required manner.

fetch protection A system-provided restriction to prevent a program from accessing data in another user's segment of storage.

file protection The aggregate of all processes and procedures in a system designed to inhibit unauthorized access, contamination, or elimination of a file.

file security The means by which access to computer files is limited to authorized users only.

flaw hypothesis methodology A systems analysis and penetration technique in which specifications and documentation for the system are analyzed and then the flaws in the system are hypothesized. The list of hypothesized flaws is then prioritized on the basis of the estimated probability that a flaw exists and, assuming a flaw does exist, on the ease of exploiting it, and on the extent of control or compromise it would provide. The prioritized list is used to direct a penetration attack against the system.

formal access approval Documented approval by a data owner to allow access to a particular category of information.

front-end security filter A security filter, which could be implemented in hardware or software, that is logically separated from the remainder of the system to protect the system's integrity.

functional testing The segment of security testing in which the advertised security mechanisms of the system are tested, under operational conditions, for correct operation.

granularity An expression of the relative size of a data object; e.g., protection at the file level is considered coarse granularity, whereas protection at the field level is considered to be of a finer granularity.

guard A processor that provides a filter between two disparate systems operating at different security levels or between a user terminal and a database to filter out data that the user is not authorized to access.

handshaking procedure A dialogue between two entities (e.g., a user and a computer, a computer and another computer, or a program and another program) for the purpose of identifying and authenticating the entities to one another.

host to front-end protocol A set of conventions governing the format and control of data that is passed from a host to a front-end machine.

identification The process that enables recognition of a entity by a system, generally by the use of unique machine-readable user names.

impersonating Synonymous with **spoofing.**

incomplete parameter checking A system design flaw that results when all the parameters have not been fully anticipated for accuracy and consistency, thus making the system vulnerable to penetration.

individual accountability The ability to associate positively the identity of a user with the time, method, and degree of access to a system.

integrity Sound, unimpaired, or perfect condition.

internal security controls Hardware, firmware, and software features within a system that restricts access of resources (hardware, software, and data) to authorized subjects only (persons, programs, or devices).

isolation The containment of subjects and objects in a system in such a way that they are separated from one another, as well as from the protection controls of the operating system.

least privilege The principle that requires that each subject be granted the most restrictive set of privileges needed for the performance of authorized tasks. The application of this principle limits the damage that can result from accident, error, or unauthorized use.

limited access Synonymous with **access control**.

list-oriented A computer protection system in which each protected object has a list of all subjects authorized to access it. Compare with **ticket-oriented.**

lock-and-key protection system A protection system that involves matching a key or password with a specific access requirement.

logic bomb A resident computer program that triggers the perpetration of an unauthorized act when particular states of the system are realized.

loophole An error of omission or oversight in software or hardware that permits circumventing the system security policy.

magnetic remanence A measure of the magnetic flux density remaining after removal of the applied magnetic force. Refers to any data remaining on magnetic storage media after removal of the power (e.g., after erasure of a diskette with a degausser).

maintenance hook Special instructions in software to allow easy maintenance and additional feature development. These are not clearly defined during access for design specification. Hooks frequently allow entry into the code at unusual points or without the usual checks, so they are a serious security risk if they are not removed prior to live implementation. Maintenance hooks are special types of trap doors.

malicious logic Hardware, software, or firmware that is intentionally included in a system for an unauthorized purpose; e.g., a Trojan Horse.

mandatory access control A means of restricting access to objects based on the sensitivity (as represented by a label) of the information contained in the objects and the formal authorization (i.e., clearance) of subjects to access information of such sensitivity. Compare with **discretionary access control.**

masquerading Synonymous with **spoofing.**

mimicking Synonymous with **spoofing.**

multilevel device A device that is used in a manner that permits it to simultaneously process data of two or more security levels without risk of compromise. To accomplish this, sensitivity labels are normally stored on the same physical medium and in the same form (i.e., machine-readable or human-readable) as the data being processed.

multilevel secure A class of system containing information with different sensitivities that simultaneously permits access by users with different security clearances and needs to know, but prevents users from obtaining access to information for which they lack authorization.

mutually suspicious The state that exists between interacting processes (subsystems or programs) in which neither process can expect the other process to function securely with respect to some property.

need to know The necessity for access to, knowledge of, or possession of specific information required to carry out official duties.

network front end A device that implements the necessary network protocols, including security-related protocols, to allow a computer system to be attached to a network.

object A passive entity that contains or receives information. Access to an object potentially implies access to the information it contains. Examples of objects are records, blocks, pages, segments, files, directories, directory trees, and programs, as well as bits, bytes, words, fields, processors, video displays, keyboards, clocks, printers, and network nodes.

object reuse The reassignment and reuse of a storage medium (i.e., page frame, disk sector, or magnetic tape) that once contained one or more objects. To be securely reused and assigned to a new subject, storage media must contain no residual data (magnetic remanence) from the object(s) previously contained in the media.

Orange Book Alternate name for *DoD Trusted Computer Security Evaluation Criteria.*

overwrite procedure A stimulation to change the state of a bit followed by a known pattern. See **magnetic remanence.**

password A protected, private character string used to authenticate an identity.

penetration The successful act of bypassing the security mechanisms of a system.

penetration signature The characteristics or identifying marks that may be produced by a penetration.

penetration study A study to determine the feasibility and methods for defeating controls of a system.

penetration testing The portion of security testing in which the evaluators attempt to circumvent the security features of a system. The evaluators may be assumed to use all system design and implementation documentation, which may include listings of system source code, manuals, and circuit diagrams. The evaluators work under the same constraints applied to ordinary users.

permissions A description of the type of authorized interactions a subject can have with an object. Examples include read, write, execute, add, modify, and delete.

personnel security The procedures established to ensure that all personnel who have access to sensitive information have the required authority as well as appropriate clearances.

physical security The application of physical barriers and control procedures as preventive measures or countermeasures against threats to resources and sensitive information.

piggyback Gaining unauthorized access to a system via another user's legitimate connection. See **between-the-lines entry.**

preferred products list (PPL) A list of commercially produced equipments that meet TEMPEST and other requirements prescribed by the National Security Agency. This list is included in the NSA Information Systems Products and Services Catalogue, issued quarterly and available through the Government Printing Office.

print suppression Eliminating the displaying of characters in order to preserve their secrecy; e.g., not displaying the characters of a password as it is keyed at the input terminal.

privileged instructions A set of instructions (e.g., interrupt handling or special computer instructions) to control features (such as storage protection features) that are generally executable only when the automated system is operating in the executive state.

procedural security Synonymous with **administrative security.**

process A program in execution. See **domain** and **subject.**

protection philosophy An informal description of the overall design of a system that delineates each of the protection mechanisms employed. A combination, appropriate to the evaluation class, of formal and informal techniques is used to show that the mechanisms are adequate to enforce the security policy.

protection ring One of a hierarchy of privileged modes of a system that gives certain access right to user programs and processes authorized to operate in a given mode.

protocols A set of rules and formats, semantic and syntactic, that permits entities to exchange information.

pseudo flaw An apparent loophole deliberately implanted in an operating system program as a trap for intruders.

Public Law 100-235 Also known as the Computer Security Act of 1987, this law created a means for establishing minimum acceptable security practices for improving the security and privacy of sensitive information in federal computer systems. This law assigns to the National Institute of Standards and Technology the responsibility for developing standards and guidelines for federal computer systems processing unclassified data. The law also requires establishment of security plans by all operators of federal computer systems that contain sensitive information.

purge The removal of sensitive data from an automated information system storage device, or peripheral device with storage capacity, at the end of a processing period. This action is performed in such a way that there is assurance proportional to the sensitivity of the data that the data may not be reconstructed. An automated information system must be disconnected from any external network before a purge. After a purge, the medium can be declassified by observing the review of procedures of the respective agency.

read A fundamental operation that results only in the flow of information from an object to a subject.

read access Permission to read information.

recovery procedures The actions necessary to restore a system's computational capability and data files after a system failure.

reliability The probability of a given system performing its mission adequately for a specified period of time under the expected operating conditions.

residual risk The portion of risk that remains after security measures have been applied.

residue Data left in storage after processing operations are complete, but before degaussing or rewriting has taken place.

restricted area Any area to which access is subject to special restrictions or controls for reasons of security or safeguarding of property or material.

risk The probability that a particular threat will exploit a particular vulnerability of the system.

risk analysis The process of identifying security risks, determining their magnitude, and identifying areas needing safeguards. Risk analysis is a part of risk management and is synonymous with **risk assessment.**

risk assessment Synonymous with **risk analysis.**

risk management The total process of identifying, controlling, and eliminating or minimizing uncertain events that may affect system resources. It includes risk analysis, cost benefit analysis, selection, implementation and test, security evaluation of safeguards, and overall security review.

safeguards See **security safeguards.**

scavenging Searching through object residue to acquire unauthorized data.

secure configuration management The set of procedures appropriate for controlling changes to a system's hardware and software structure for the purpose of ensuring that changes will not lead to violations of the system's security policy.

secure state A condition in which no subject can access any object in an unauthorized manner.

security critical mechanisms Those security mechanisms which must operate correctly to ensure that the security policy is enforced.

security evaluation An evaluation done to assess the degree of trust that can be placed in systems for the secure handling of sensitive information. One type, a product evaluation, is an evaluation performed on the hardware and software features and assurances of a computer product from a perspective that excludes the application environment. The other type, a system evaluation, is done for the purpose of assessing a system's security safeguards with respect to a specific operational mission and is a major step in the certification and accreditation process.

security features The security-relevant functions, mechanisms, and characteristics of system hardware and software. Security features are a subset of system security safeguards.

security filter A trusted subsystem that enforces a security policy on the data that pass through it.

security flaw An error of commission of omission in a system that may allow protection mechanisms to be bypassed.

security flow analysis A security analysis performed on a formal system specification that locates potential flows of information within the system.

security label A piece of information that represents the security level of an object.

security level The combination of a hierarchical classification and a set of nonhierarchical categories that represents the sensitivity of information.

security measures Elements of software, firmware, hardware, or procedures that are included in a system for the satisfaction of security specifications.

security perimeter The boundary where security controls are in effect to protect assets.

security policy The set of laws, rules, and practices that regulate how an organization manages, protects, and distributes sensitive information.

security range The highest and lowest security levels that are permitted in or on a system, system component, subsystem, or network.

security requirements The types and levels of protection necessary for equipment, data, information, applications, and facilities to meet the needs of a security policy.

security requirements baseline A description of minimum requirements necessary for a system to maintain an acceptable level of security.

security safeguards The protective measures and controls that are prescribed to meet the security requirements specified for a system. Those safeguards may include but are not necessarily limited to hardware and software security features, operating procedures, accountability procedures, access and distribution controls, management constraints, personnel security, and physical structures, areas, and devices. Also called **safeguards.**

security specifications A detailed description of the safeguards required to protect a system.

security test and evaluation An examination and analysis of the security safeguards of a system as they have been applied in an operational environment to determine the security posture of the system.

security testing A process used to determine that the security features of a system are implemented as designed. This includes hands-on functional testing, penetration testing, and verification.

sensitive information Any information, the loss, misuse, modification of, or unauthorized access to, that could affect an organization.

single-level device An automated information systems device that is used to process data of a single security level at any one time.

social engineering The process of subverting security measures by taking advantage of an individual in order to learn passwords or other security information without resorting to a technical attack.

software security General purpose (executive, utility, or software development tools) and applications programs or routines that protect data handled by a system.

software system test and evaluation process A process that plans, develops, and documents the quantitative demonstration of the fulfillment of all baseline functional performance, operational, and interface requirements.

spoofing An attempt to gain access to a system by posing as an authorized user. Synonymous with **impersonating, masquerading,** or **mimicking.**

stand-alone, shared system A system that is physically and electrically isolated from all other systems, and is intended to be used by more than one person, either simultaneously (e.g., a system with multiple terminals) or serially, with data belonging to one user remaining available to the system while another user is using the system (e.g., a personal computer with nonremovable storage media such as a hard disk).

stand-alone, single-user system A system that is physically and electrically isolated from all other systems, and is intended to be used by one person at a time, with no data belonging to other users remaining in the system (e.g., a personal computer with removable storage media such as a floppy disk).

subject An active entity, generally a person, process, or device, that causes information to flow among objects or changes the system state.

system integrity The quality that a system has when it performs its intended function in an unimpaired manner, free from deliberate or inadvertent unauthorized manipulation of the system.

tampering An unauthorized modification that alters the proper functioning of an equipment or system in a manner that degrades the security or functionality it provides.

technical attack An attack that can be perpetrated by circumventing or nullifying hardware and software protection mechanisms, rather than by subverting system personnel or other users.

technical vulnerability A hardware, firmware, communication, or software flaw that leaves a computer processing system open for potential exploitation, either externally or internally, thereby resulting in risk for the owner, user, or manager of the system.

TEMPEST The study and control of spurious electronic signals emitted by electrical equipment.

terminal identification The means used to uniquely identify a terminal to a system.

threat Any circumstance or event with the potential to cause harm to a system in the form of destruction, disclosure, modification of data, and/or denial of service.

threat agent A method used to exploit a vulnerability in a system, operation, or facility.

threat analysis The examination of all actions and events that might adversely affect a system or operation.

threat monitoring The analysis, assessment, and review of audit trails and other data collected for the purpose of searching out system events that may constitute violations or attempted violations of system security.

threat multiplier An entity whose involvement magnifies a threat.

ticket-oriented A computer protection system in which each subject maintains a list of unforgeable bit patterns, called tickets, one for each object the subject is authorized to access. Kerberos is a ticket-oriented protection system. Compare with **list-oriented.**

time-dependent password A password that is valid only at a certain time of day or during a specified interval of time.

trap door A hidden software or hardware mechanism that can be triggered to permit system protection mechanisms to be circumvented. It is activated in some innocent-appearing manner; e.g., a special "random" key sequence at a terminal. Software developers often introduce trap doors in their code to enable them to reenter the system and perform certain functions. Synonymous with **back door.**

Trojan Horse A computer program with an apparently or actually useful function that contains additional (hidden) functions that surreptitiously exploit the legitimate authorizations of the invoking process to the detriment of security or integrity.

trusted computer system A system that employs sufficient hardware assurance measures to allow its use for simultaneous processing of a range of sensitive or classified information.

trusted identification forwarding An identification method used in networks whereby the sending host can verify that an authorized user on its system is attempting a connection to another host. The sending host transmits the required user authentication information to the receiving host. The receiving host can then verify that the user is validated for access to its system. This operation may be transparent to the user. See also **advanced authentication.**

trusted process A process, the incorrect or malicious execution of which is capable of violating system security policy.

untrusted process A process that has not been evaluated or examined for adherence to the security policy. It may include an incorrect or a malicious code that attempts to circumvent the security mechanisms.

user A person or process accessing an automated information system either by direct connections (i.e., via terminals), or indirect connections (i.e., prepare input data or receive output that is not reviewed for content or classification by a responsible individual).

user ID A unique symbol or character string that is used by a system to identify a specific user.

user profile Patterns of a user's activity that can be used to detect changes in normal routines.

verification The process of comparing two levels of system specification for proper correspondence (e.g., security policy model with a top-level specification, top-level specification with source code, or source code with object code). This process may or may not be automated.

virus A self-propagating Trojan Horse, composed of a mission component, a trigger component, and a self-propagating component.

vulnerability A weakness in system security procedures, system design, implementation, internal controls, etc., that could be exploited to violate system security policy.

vulnerability analysis The systematic examination of systems in order to determine the adequacy of security measures, identify security deficiencies, and provide data from which to predict the effectiveness of proposed security measures.

vulnerability assessment A measure of vulnerability that includes the susceptibility of a particular system to a specific attack and the opportunities available to a threat agent to mount that attack.

work factor An estimate of the effort or time needed by a potential penetrator with specified expertise and resources to overcome a protective measure.

write A fundamental operation that results only in the flow of information from a subject to an object.

write access Permission to write to an object.

Resources

AT&T, *BCSystem Security Handbook*. A 70-page abridgement of John Hough's book. Available for a small fee from AT&T General Business Systems Division.

AT&T, *10 Tips to Help Secure Your Network,* free on request.

AT&T Corporate Security Seminars held quarterly by region.

Auerbach Publications, *Information Systems Security*, Boston, MA. A quarterly magazine for security professionals (800) 950-1205, $130/year. Also from Auerbach: *EDPACS Security Newsletter* and *Data Security Management.*

Cooper, James Arlin, *Computer and Communications Security,* McGraw-Hill, New York, 1989.

Hough, John, *Toll Fraud And Telabuse: A Multibillion Dollar National Problem,* Telecommunications Advisors, Inc., Portland, OR. Also available through NCSA.

Institute of Electrical and Electronics Engineers, *IEEE 802.10 Interoperable LAN/MAN Security (SILS),* 445 Hoes Lane, P.O. Box 1331, Piscataway, NJ 08855-1331, (908)562-5423.

National Computer Security Center, INFOSEC Awareness Division, Attn: X711/IAOC, Fort George G. Meade, MD 20755-6000, (410) 766-8729 (Barbara Keller). Multicolored (Red, Green, Orange, etc.) Security Books from NSA, the industry standard.

National Technical Information Service, source of FIPS Standards, 5285 Port Royal Road, Springfield, VA 22161, (703) 487-4650.

NCSA Buyer's Guide, 10 South Courthouse Avenue, Carlisle, PA 17013. Computer Security Resource Database (via Network World): (508) 620-1160 8N1 300 to 2,400 bits/s; (508) 620-1178 8N1 300 to 9,600 bits/s.

North American Telecommunications Association, *Remote Access Toll Fraud: Detection and Protection,* Dept. 5137, Washington, D.C., 20061-5137, (800) 538-6282 ext. 260, $53 for nonmembers or $33 for NATA members. This is a 52-page report to help fend off PBX violators.

Spafford, Heaphy, and Ferbrache, *Computer Viruses,* ADAPSO (the computer software and services industry association). A thorough description of computer virus types, antiviral techniques, and legal issues. Contains names and addresses of many further resources and reading.

Stang, David, Ph.D. and Sylvia Moon, *Network Security Secrets,* IDG Books, International Computer Security Association, Washington, D.C.

Wood, Charles Cresson, *Information Security Policies Made Easy,* available through NCSA, 525 Information Security Policies.

Articles

Adelson, Andrea, "Software Sounds the Alarm on Cellular Theft," *The New York Times,* July 28 (1993), Business Technology section. Description of Coral, GTE, and EDS/PacTel cellular security software for cellular carriers.

Buchsbaum, S. J., "Networked Computer Security; Processes and Technology," *Vital Speeches,* **58,** December 15: 150 (1991) (transcript of an address).

Business Week, October 28: 134A (1991). Brief description of Moscom Corp.'s PBX VoiceLock product.

Daly, James, "Defense Firms Eye Commercial Security; Hughes STX, ICL Join Forces to Provide Enterprisewide Net Protection," *Computerworld,* **26,** July 13: 53 (1992).

Daly, James, "Federal Computer Systems Still Vulnerable; Report Warns of Internet-Type Disaster if Government Underestimates Need for Security," *Computerworld,* **26,** June 8: 55 (1992). Input (consulting and research firm), compiled a 160-page report on the federal security market.

Goggins, Chris, "Hackers Aren't the Real Enemy," *Computerworld,* **26,** June 8: 37 (1992). Discussion of hacker motivations and psychological profile.

Kenner, Hugh, "Stomping the Nasties," *Byte,* **15,** November: 466 (1990). A guide to top computer viruses.

Mamis, Robert A., "Fending off Phone Fraud," *Inc.,* **15,** May: 45 (1993). Description of Dodger Industries, Inc. $35,000 loss.

Port, Otis, "An early-warning system against computer invaders," *Business Week,* December 2: 140 (1991). Discusses protection systems developed by Lawrence Livermore National Laboratory.

Vollmart, Sarah, "How to Implement a PC Security System," *The Office,* **115** February: 42 (1992). Discusses protection techniques from authentication to audits. Recommends an enterprise approach toward security managements and standards.

Wallich, Paul, "Safe Bytes: What It Will Take to Keep Computers Secure," *Scientific American* (Science and the Citizen) **264,** February: 28 (1991).

Index

ABOUT THE AUTHOR

Fred Simonds has more than 19 years of experience in the field of communications. Based in Trumbull, Connecticut, he consults on behalf of firms who seek to build, secure, or reengineer their data and voice networks. Mr. Simonds is a nationally known teacher and speaker on communications topics and is an adjunct faculty member at Pace University in Westchester, New York. Formerly, Mr. Simonds spent four years working in hardware design and systems engineering at Motorola Codex and nine years in applications engineering with Racal-Datacom. His military experience includes six years of secure networking using NSA-supplied COMSEC equipment. He is the author of *The McGraw-Hill LAN Communications Handbook*, a main selection of The Library of Computer and Information Sciences.